Science from Your Airplane Window

Science from Your Airplane Window

Elizabeth A. Wood

Illustrated with Photographs and with Drawings by Frank M. Thayer

SECOND REVISED EDITION

DOVER PUBLICATIONS, INC., NEW YORK

Published in Canada by General Publishing Company, Ltd., 30 Lesmill Road, Don Mills, Toronto, Ontario.
Published in the United Kingdom by Constable and Company, Ltd., 10 Orange Street, London WC 2.

This Dover edition, first published in 1975, is an unabridged and revised republication of the work first published in 1968 by Houghton Mifflin Company under the title *Science for the Airplane Passenger*.

International Standard Book Number: 0-486-23205-0
Library of Congress Catalog Card Number: 75-14762

Manufactured in the United States of America
Dover Publications, Inc.
180 Varick Street
New York, N.Y. 10014

Acknowledgments

IT WAS at a conference in Boulder, Colorado, in the summer of 1964 that I said to Professor Gerald Holton of Harvard, "Someone should write a book about all the wonderful things one sees from the air, especially as they illustrate the principles of science." "Why don't you do it?" he said to me, and that was how this book began.

A great many people have contributed to it. Gerald Holton has kept in close touch with it at every stage. With Robert Chesley and other members of the staff of Harvard Project Physics he has offered valuable constructive criticism of the manuscript. The administration of Bell Telephone Laboratories has encouraged me to take time to write the book because of its educational value, not only to the student of science, but also to the intelligent citizen who is curious about the world around him, above him and below him.

When a book includes parts of several fields of science, as this book does, the author must either be skilled in all these fields or must seek help from others. My training is in geology and physics and I have depended on the rich background of many friends where my own background was inadequate.

Sidney Millman, David Slepian, Sidney Darlington, Bernard

Stevens and Kurt Nassau, physicists, mathematicians and chemists at Bell Laboratories, read the whole book and gave me very helpful detailed criticism. Mrs. Sidney Darlington also did this for me, from the point of view of the interested layman. Kenneth McKay, Bernd Matthias and John Klauder, of Bell Laboratories, and Arnold Strassenburg, of the State University of New York at Stony Brook, added air observations and performed experiments in flight for me. Others at Bell Laboratories who have contributed to the book are Richard Haynes, Vera Compton, Jeff Courtney-Pratt, Larner Gambrill, Alan Holden, Irene Longley, J. W. Schaefer, Frank Sinden and Willem Van Bergeijk, who knows about ears.

Arthur Strahler, of the Geology Department of Columbia University, reviewed the geological chapters, and Pierre Dansereau, of the New York Botanical Garden, helped me on matters related to vegetation. Mary Elizabeth Solari of the Mathematics Department of the Chelsea College of Science and Technology in London, provided the geometrical proof for the measurement of the distance between window panes and called my attention to many things best seen from the air. Lincoln Dryden, of the Geology Department of Bryn Mawr College, Frank Solari, of the Ministry of Aviation, Great Britain, and Gene Weltfish, anthropologist, of Fairleigh Dickinson University, helped me find the literature on archaeology from the air.

One of the pleasures of writing this book has been making new friends because of the need to consult experts in various areas. Time and again I came to such people as a complete stranger, offering nothing but my need to know what they could tell me. In the midst of their busy lives they took time to discuss with me the questions to which I needed answers. Among these I want especially to mention the meteorologists, F. H. Ludlam of the Imperial College of Science and Technology of the University of London, Hans Neuberger of Pennsylvania State University, and Vincent J. Schaefer, Director of the Atmospheric Sciences Research Center of the State University of New York at Albany. The art of spiral plowing was elucidated for me by Truman May and Robert Hood, Farm Advisers of Madison and St. Clair Counties, Illinois, and by Frank Andrew of the University of Illinois who invented the method of doing it "no hands."

In matters related to the construction and handling of the plane I have turned to Lewis Larmore of Douglas Aircraft, Captain R. T. Dawe of United Airlines, and R. V. Brady of Sperry Gyroscope. It was Robert Dixon, Jr., the assistant librarian of the U.S. Coast Guard Academy at New London, Connecticut, who gave me the historical data on port and starboard lights; and L. E. Stone, of the Federal Aviation Agency at Newark Airport, who told me about airport lights and regulations.

All of these are, in a sense, co-authors, but they are not to be held responsible for those weaknesses or errors which may have crept into the text after they saw it.

The drawings throughout the book were done by Frank M. Thayer of Bell Telephone Laboratories, for whose interest, skill and artistic talent I am indeed grateful. Ann S. Cooper of Bell Telephone Laboratories proofread the entire book and handled many end-stage matters for me.

Every author hopes for a sympathetic publisher. The personal interest and enthusiasm of Paul Brooks, Editor-in-Chief and Director of Houghton Mifflin Company, and of David Harris, who has worked closely with me on the book, have made its publication a pleasure.

Finally I want to express my deep appreciation of the thoughtfulness and understanding of my husband, Ira Wood. Without his cooperation the book could not have been written.

Preface

THIS BOOK was written for people interested in science. You don't have to have had a course in science to read it. Neither do you have to be a student. It was written for fun, to be read for fun.

The use of our senses to learn about the world around us and the use of our minds to wonder about it are at the very heart of all science. A plane trip is an opportunity to see and hear and feel things that we don't experience every day. The purpose of this book is to help you make the most of this opportunity. It need not be read consecutively. Skip around to sections that interest you at the moment.

You will find little in it about the airplane itself because there are so many different types of planes. It seemed better to leave the discussion of the airplane's internal workings to the leaflets written by the airplane companies. Some of them do it very well.

Many of the observations mentioned in this book were first called to my attention by others: my observing friends, people I met on planes, and of course by other authors in whose books I discovered clues to interesting things to watch for. You, the reader, will discover other things which should have been included. I hope you will write and tell me about them.

ELIZABETH A. WOOD

Contents

Illustrations

Color Plates

Tables

Science from Your Airplane Window

1. *Taking Off*

THE PLANE is on the runway. You are comfortably seated with your seat belt fastened according to instructions and you have opened this book. Here we go.

The plane surges forward along the runway. But what happens to you? You go along, reluctantly. Your body hangs back, pushing against the back of the chair. If your body were not there, that force would not be exerted in a backward direction on the chair. The engine has to work just that much harder to accelerate the plane because you are on board.

The way your body hangs back demonstrates a property that every body has: inertia. Inertia is the tendency of a body to continue standing still if it is standing still or to continue moving at constant velocity in a straight line if it is moving at constant velocity in a straight line. As the plane accelerates, increases its velocity, along the runway, the force between your body and the chair continues to be felt because your body tends to maintain a constant velocity unless a force is exerted upon it.

When the speed is great enough, the pilot decides to take off. You feel a bit of a lift and you are in the air. What do you mean, you "feel a lift"? You would have known with your eyes closed that the plane was taking off. How? Your body was speeding down the runway in a straight line when the path of the plane curved upward from the ground. The inertia of your body made it tend to continue in a straight line so again you and the plane pushed on each other. In the early days of flying, when few pilots had instruments to tell them whether the plane was rising or falling, those without instruments were said to be "flying by the seat of their pants."

Suppose you were weighing yourself on a scale in the airplane instead of sitting in your seat when the plane left the ground. The scale would exert an additional upward force on you, just as the seat did. This is the same as saying that you would exert an additional downward force on the scale: you and the scale would push against each other. If you read your weight on the scale you would find it to be greater than it was when the plane was not accelerating upward. If you found it to be one tenth greater, you could say that you were experiencing an acceleration of 0.1 g because your weight was 1.1 times what it would have been if due just to the gravitational attraction between you and the earth. The g stands for gravitational acceleration, 32 ft. per second per second, of a body falling in vacuum. In air force tests men have survived accelerations of more than 30 g without permanent injury.

I have never seen a people scale on an airplane, but you can carry a letter scale on board with you and measure vertical acceleration by weighing something on it. You may be surprised at how small the acceleration is when you can feel it distinctly. A qualitative measure can be obtained from the difference in stretch of a rubber band with a heavy object, such as a bunch of keys, suspended from it. The difference will be small so you will need to hold something behind the rubber band against which you can judge the change in position of the suspended weight.

Sir Isaac Newton would have enjoyed the experience of taking off. Back in 1685 he stated precisely what had previously been noticed by Galileo: "Every body perseveres in its state of rest, or of uniform motion in a straight line, unless it is compelled to change that state by forces impressed thereon."* There it is, the pressure on your back when your "body perseveres in its state of rest"; the pressure on the seat of your pants "by forces impressed thereon" when your "body perseveres in its state . . . of uniform motion in a straight line." This is known as Newton's first law of motion.

Sometimes there are things worth watching for as you go down the runway. If there is snow on the ground, the snow near the runway may become covered with small icy snowballs, rolled up by the strong air blasts from the passing planes.

If there are puddles of water the blast of air from the plane causes their surfaces to be covered with ripples. In the same way the wind causes the sea to be covered with waves, but here at the airport the phenomenon is miniaturized, both in space and time. Hundreds of very small waves form in a very short time. Why doesn't the air pushing on the surface of the water just push a hollow in it? Why, instead, does it cause many evenly spaced troughs with ridges in between? The distance from ridge to ridge is very regular. What does it depend on? The answers to some of these questions are difficult to determine, even for students of hydrodynamics.

The stable position of the surface of standing water is a horizontal plane, perpendicular to the gravitational force between it and the earth. If standing water is disturbed, as it is when a stone is dropped into it, waves form. Water is a pretty incompressible substance. If you push down on its surface in one place, the surface will have to rise somewhere else. In

* Isaac Newton, *Principia*, 2-volume paperback (Berkeley, University of California Press, 1962).

both places the surface will then no longer be in the stable position and will be subject to gravitational forces tending to restore the original horizontal plane surface. But the surface in each case "overshoots" the stable position, the trough becomes a crest and the crest becomes a trough. The wave form moves on from one place to the next as the water oscillates up and down. We have not answered all the questions we asked. The purpose of asking them is to call attention to the extraordinariness of ordinary things.

As the plane takes off, it loses contact with its shadow. Whatever flies gets separated from its shadow.

Airplanes take off into the wind. At first this might seem strange since the plane has to push against the wind and it is trying to gain speed. A racing car can't go as fast against the wind as with the wind and neither can a plane. However, the important thing in getting an airplane off the ground is not its speed relative to the ground, but its speed relative to the air. If it can get up to 100 mph. with a 25 mph. head wind, its speed relative to the air is 125 mph. It could go faster, relative to the ground, down wind — perhaps 110 mph. But then its speed relative to the air would be only 85 mph.

If the speed of the plane relative to the air makes it rise, it must be that the forward motion causes more up-push than down-push, a net force upward. The air must somehow be pressing harder against the underside of the wings than against the top side, so much harder that it lifts the airplane. One way in which this difference in pressure is achieved is by shaping the wing so that the air that goes over the top has to go farther than the air that goes under the bottom (Fig. 1). As with soup, when the same amount has to go farther, it has to be thinner. The molecules of air passing over the top of the wing spread farther apart and exert less pressure than those underneath the wing where they are packed more closely together. At the trailing edge of the wing the top air must come together again with the bottom air. The greater velocity of the air going over the top has resulted in a lifting force.

You can demonstrate this effect with a small piece of paper. One about 3 × 5 inches is satisfactory. Hold the short edge close beneath your lower lip and blow across the top of the paper (Fig. 2). The greater velocity of the air above the paper makes it rise. It is just this that helps to lift the plane off the ground and keep it in the air.

Figure 1. Air foil

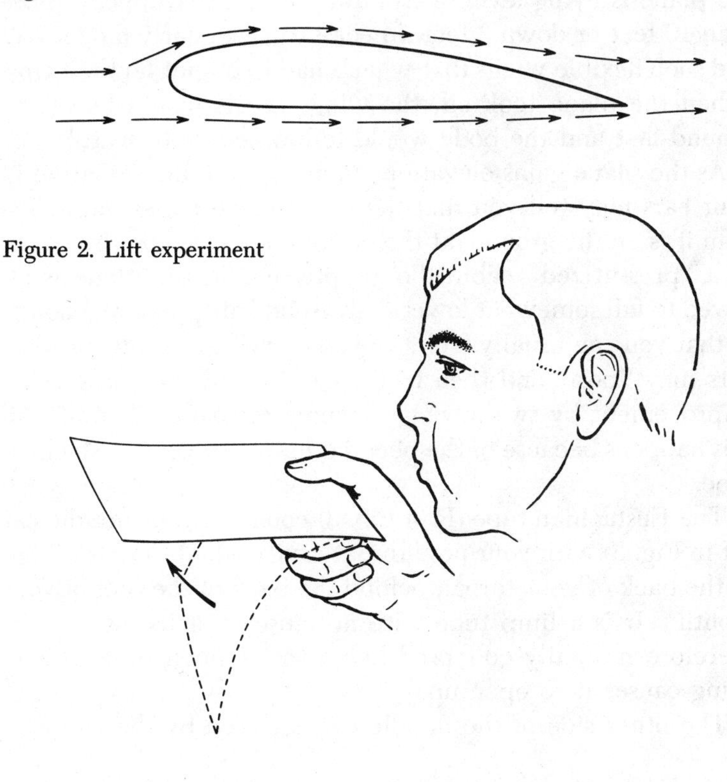

Figure 2. Lift experiment

The air moving past the wing helps to lift the plane in still another way. If the leading edge of the wing is higher than the trailing edge, the force of the moving air against the under side of the wing will be greater than its force against the upper side of the wing. The lift produced in this way has been experienced by everyone who has held his hand, with fingertips lifted forward, in the airstream passing the window of a fast-moving car.

In the air the body of the plane hangs between the air-supported wings which bend because of their burden. The outer tip of the wing of a Boeing 707 is raised 39 inches when the plane is flying level in still air, but can flex upward more than 10 feet or down 3 feet if necessary. An early model B47 had such flexible wings that wheels had to be put on their tips. When the plane took off, the wing wheels would leave the ground first and the body would follow soon afterward!

As the plane gains elevation, an uncomfortable sensation in your ears may tell you that the air pressure is becoming less than it is on the ground, at the bottom of the sea of air. Even in a "pressurized" cabin the air pressure in the plane is allowed to fall somewhat lower than ground air pressure, enough so that you can usually feel it in your ears. After a while your ears may "pop" and then feel better or you may hasten the improvement by swallowing, yawning or chewing gum. All this happens because of the peculiar nature of your Eustachian tube.

The Eustachian tube (E in Fig. 3) connects your middle ear (M in Fig. 3) with your nose-throat cavity which it enters high in the back of your throat, behind the back of the roof of your mouth. It is a limp tube with no muscles of its own and is therefore normally collapsed like a toy balloon unless something causes it to open up.

The other side of the middle ear is closed by the eardrum

(D in Fig. 3), which seals off the middle ear from the outside air. The middle ear is about half an inch in length from front to back and half an inch high — and it is full of air.

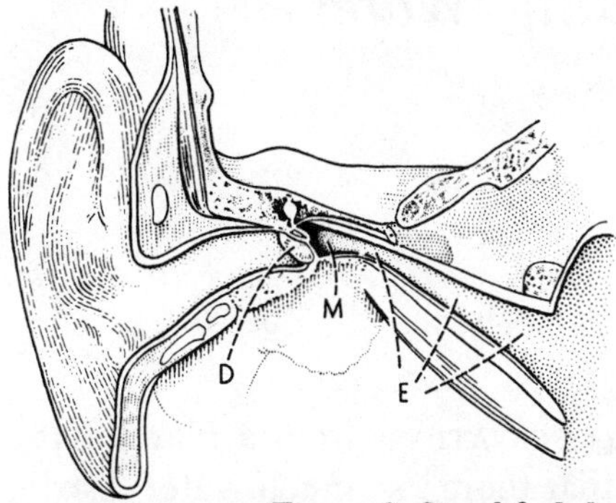

Figure 3. Simplified diagram of the ear

On the ground, the air in the middle ear is at ground-level air pressure. As the outside pressure is lowered, the trapped air in the middle ear presses outward on the surrounding tissues, giving you the sensation of pressure in your ears. The pressure on your eardrum distorts it and your hearing is somewhat impaired. When the difference between the inside pressure and the outside pressure becomes great enough to push the walls of the Eustachian tube outward, it is forced open (your ears pop) and some air is released from the middle ear so that the pressures are equalized. Sometimes the escape of air is apparently achieved as a slow leak along the Eustachian tube, since your ears gradually get to feeling better without popping. When you swallow or yawn, your muscles pull on the end of the Eustachian tube, straightening it out so that it can be forced open more easily than when it is crumpled.

We will come back to physical sensations later. Now let us see what we can observe from the window.

2. *Standing Water*

ONE OF THE OBSERVATIONS from a plane that is surprising to most of us is that there's so much water distributed across the land. Over much of the northern United States and nearly all of Canada there is a great abundance of lakes and streams, most of which we never see from the ground because, though our road may be close to them, vegetation or some slight rise of land hides them from view.

If you are on the sunlit side of the plane, watch for the water that is just in the right position to reflect the sun's rays into your eye. When the sun is high in the sky the brightly reflecting water will be close below the plane; when the sun is low, it will be far off. This is because rays that reach the lake's surface at a steep angle will also leave at a steep angle, but when they arrive more nearly parallel to the surface, they leave more nearly parallel and you would have to be flying very low to intercept them if you were close to the lake. It is the law of reflection: "The angle of incidence is equal to the angle of reflection." (A = A, B = B, Fig. 4) Every little puddle and rivulet glitters brightly as it reflects the sun and you can then appreciate how much water there really is (Plate I).

Plate I. The water in the right position flares into brilliance

If the water has waves or ripples on it so that parts of its surface are tilted away from the horizontal, these parts will reflect light at an angle different from that for the horizontal surface. If the still water is reflecting the sun brightly to your eye, the part of the surface that is rippled by the wind will be at the wrong angle to reflect and will look darker. If the standing water is not reflecting the direct sunlight, some of the surfaces of the waves may be at the correct angle to reflect and will look brighter.

A dark river far below you may show a straight line across it with brightly sparkling water on one side of the straight line: a dam with rough water downstream.

When bright reflections of the sunlight flash out from other parts of the landscape (other than that where the horizontal water surfaces are brightly reflecting), they must be from non-horizontal surfaces that happen to be at just the proper angle of tilt: the roof of a greenhouse, perhaps, or a skylight.

Sometimes, when the sun is low and the distant landscape is obscured by haze, a bright reflection may show you a body of water much farther away than any other feature in view. Because the water is separated in viewing angle from the rest of the landscape that you can see (see Fig. 5), it gives the impression of being higher than it ought to be, sometimes even appearing to be floating above the land.

Physicists have made for themselves mental pictures of what light is like. The picture that fits much of the behavior of light is that it is wave-like in nature and we find we can use the patterns of wave forms moving across the surface of water to help us think about some of the experimental results with light. The distance from crest to crest of neighboring waves is called the wavelength. The wavelengths of light waves are very short: there are about 36,000 wavelengths of red light to the inch. Orange, yellow, green and blue light have progressively shorter wavelengths and the wavelength of violet light is just about half as long as that of red light.

You may remember that one effect that makes us think of light rays as wave-like in nature rather than bullet-like is their behavior when they pass through a pinhole. Light from a distant source passing through a pinhole does not continue as a single rod-like beam with sharp edges which would make a perfectly sharp circular spot of light on a white piece of paper

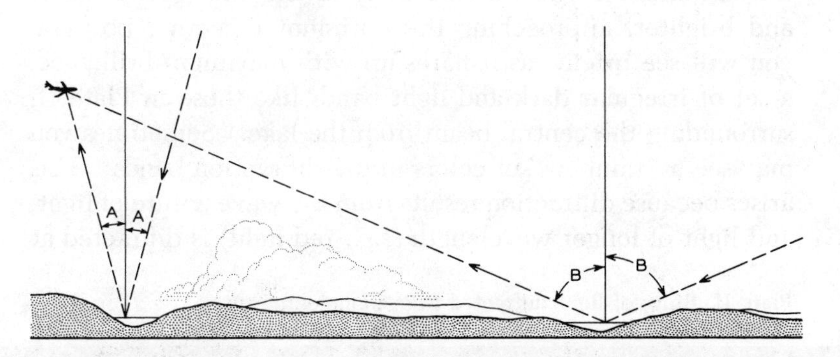

Figure 4. The angle of incidence is equal to the angle of reflection

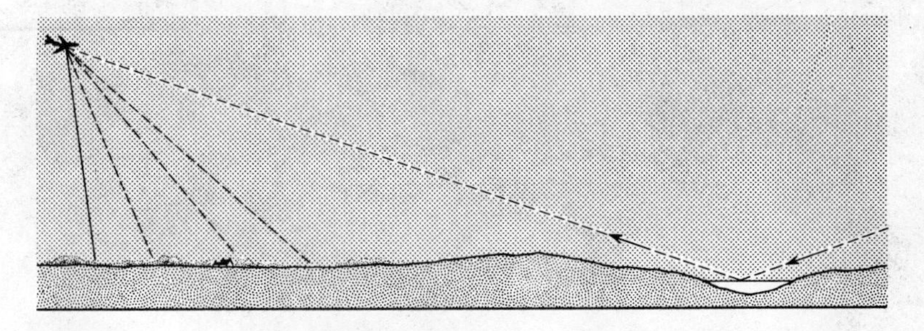

Figure 5. Distant reflecting bodies of water may appear to be floating above land

held in its path. Instead the narrow bundle of rays spreads a bit and, because of interference among them, dark and light bands appear surrounding the central beam (Plate II). The phenomenon is known as diffraction.

You can use a little lake in the same way! A very small lake, far below you, in just the right position to reflect the sun's rays most brightly brings to your eye a narrow bundle of rays from the sun. In this way it behaves as though it were a pinhole, letting through the sunlight from a long way off on the far side. If you watch closely as the lake gets brighter and brighter, approaching the optimum reflecting position, you will see briefly, as it flares up with maximum brilliance, a set of irregular dark and light bands like those in Plate II, surrounding the central beam from the lake. Sometimes you may see a "rainbow" of colors in the diffraction bands. This arises because diffraction results from the wave nature of light, and light of longer wavelength (*e.g.*, red light) is diffracted at

Plate II. Rings of light diffracted from pinhole apertures

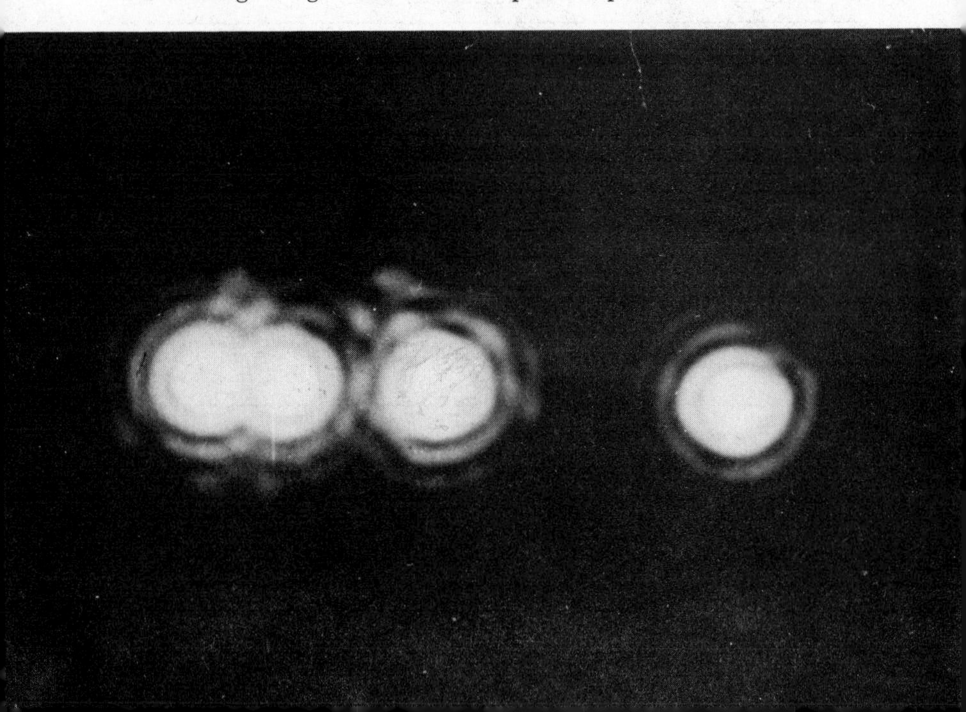

a greater angle to the central beam than is light of shorter wavelength (*e.g.*, blue light).

Light reflected from the surface of water is mostly polarized. The way in which polarized light differs from ordinary light is often illustrated in textbooks by a simple analogy with oscillations along a rope. If you hold one end of a rope with the other end tied to something and wiggle it back and forth in any direction, the oscillations will travel along the rope in a way that illustrates the travel of water waves or light waves. The form of the wave travels from one end of the rope to the other, but the rope doesn't go anywhere. If you wiggle your end of the rope sideways, sideways waves will travel down the rope; if you wiggle it from upper left to lower right, that will be the direction in which the waves oscillate along the rope.

But now suppose that between you and the far (tied) end there is a picket fence through which the rope passes. Regardless of the nature of the oscillations you introduce at your end, only those oscillations parallel to the slot in the fence will get through. On the far side of the fence the oscillations will all be in the vertical plane; that is, they will be polarized. Polarized light is light whose oscillations are all in the same plane. In ordinary light the oscillations are every which way.

When you are looking at light reflected from the surface of a body of water, the oscillations of the light waves run left and right in most of the light coming to your eye. (The physicist would say the electric vector of the electromagnetic radiation is parallel to the reflecting surface and perpendicular to the direction in which the beam is traveling.) If you have Polaroid glasses with you, you can test the reflected light for polarization by holding them in the path of the light and rotating them in their own plane (the plane of the paper in Fig. 6). The glasses let light through only when it is oscillating in a particular direction, shown by the double-ended arrows in Figure 6. They are made this way so as to cut out the horizontally oscillating light from horizontal reflecting surfaces. When the double-ended arrows are vertical there is no left-right component to their orientation at all, so none of the left-right oscillating light gets through. (Actually a little gets through because neither the Polaroid nor the water does a perfect job.) As you rotate them they come closer and closer to the left-right orientation and therefore let through more and more of the reflected light until, when the double-ended arrows are horizontal, you get maximum transmission.

As you perform this experiment from the window of the airplane, you may see colors. Further experimentation will con-

Figure 6. Rotating Polaroid glasses in their own plane

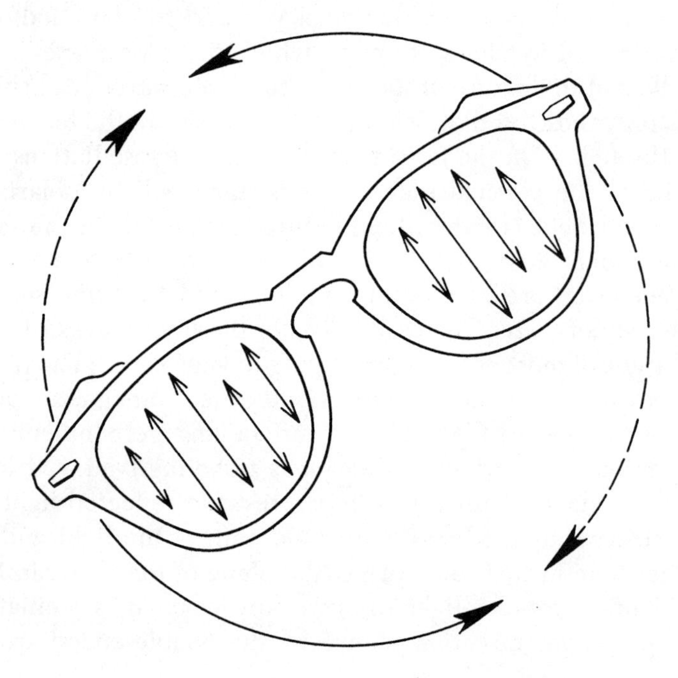

vince you that the colors are associated with the window of the airplane. Light from the clear blue sky is also polarized, as you can tell by testing it with your Polaroid glasses. When the glasses are in the right orientation to let the least light through, the center of the window will look dark, but near the edges where it is clamped by its frame it may show colors. The reason for this is that although the unstrained plastic does not affect the direction of oscillation of light, strained plastic does, and the plastic of the windows is strained by being clamped. A bit of cellophane (often furnished on the plane as wrapping for crackers or cigarettes) held between the clear blue sky and your Polaroid glasses will also show colors. The effect is even more spectacular if the cellophane is held between two pieces of Polaroid with their polarization directions at right angles to each other. The process of forming the cellophane sheet lines up the molecules in the cellophane just as the strain due to clamping lines them up at the edge of the plastic window.

The uniform circular arrangement of colors observable in this way in the whole central part of some windows (Color Plate IV) and the manner in which the color rings shift with angle of view suggest that the effect may be due, not to inhomogeneous strain resulting from deformation of the window, but to the arrangement of long molecules in the plastic perpendicular to the major surfaces of the window. In such windows, when one's angle of sight is perpendicular to the window surfaces, one sees a centered dark bull's eye surrounded by colored rings with the familiar interference-color sequence: yellow, red, blue, green, yellow, red, etc., as seen in all optically uniaxial crystals viewed along the optic axis between crossed polarizers.*

* See Plate V of *Crystals and Light* by Elizabeth A. Wood (Princeton, N.J., D. Van Nostrand Co., Inc., 1964).

When the plane climbs to regions of lower air pressure, the difference in pressure inside and outside must cause the windows to bulge a little. Then color, due to the strain, can be seen far out in the window, not just around the clamping screws. Sometimes such widespread strain is also observed in the windows when the plane is on the ground.

The optical effects of strain in plastic have made possible the study of strain in all sorts of stressed structures. Models of the structures to be studied are made of a transparent variety of Bakelite or other plastic and placed between crossed polarizers with a light behind them. When the stress is applied, the resulting light patterns show its effect on the structure (Plate III).

Plate III. Stress pattern of a large Bakelite plate compressed by a Bakelite die, showing stress concentrations at ends of die

If you fly low enough over water to see the wakes of boats you will see how two sets of waves can cross each other and continue undisturbed on their way. This is one of the extraordinary properties of all kinds of waves: light waves, sound waves and water waves. How awkward it would be if it were otherwise! Opposite pairs of people at a table for four could not communicate simultaneously. And we could never see anything because of interference from the cross fire of light beams crossing the direction in which we were looking.

Ripple tanks are tabletop boxes full of water used in laboratories for the study of waves. As the example of crossing wakes shows, lakes are giant ripple tanks when viewed from the air and phenomena demonstrated in the laboratory can be seen in them.

In some parts of some planes you may experience vibration. If so, you have a chance to observe beautiful wave phenomena in a very small ripple tank. Rest your cup of coffee on the arm of your chair. The vibration of the plane will set up standing waves on the surface of the coffee. The standing waves result from the interaction between the waves sent into the coffee by the vibration of the cup and the waves reflected back again into the coffee from the inside of the cup. Their pattern may remain the same for a few moments and then change as the vibrations of the plane change in response to air conditions and adjustments made by the crew. Place the cup so that it is in contact with the wall of the airplane and again observe the changing patterns as the pulse of the plane changes.

What color is a lake? That depends in part on what has happened to the light that comes to your eye from the lake, and that, in turn, depends in part on the angle at which you are viewing its surface. In general the proportion of light reflected from the surface is less the more nearly you are looking directly down into the lake. Lakes directly below you look dark because most of the light coming to your eye comes from below the surface, from the darker depths of the lake. Little light is reflected at such a steep angle, for most of it enters the lake. Table 1 shows the proportion of light that is reflected for various angles of incidence. The rest of it enters the lake. There it may be reflected from various objects, and if it reaches our eyes it gives us information about them. Some of the light is "absorbed" by the water and the objects in the lake. Its energy is used to cause that small rapid motion of their constituent particles that we call thermal motion. The objects become warmer.

If you fly over a large body of water, you will see how its color depends on the angle at which you view it. (See Plate VIII, for example.) Note how dark it is below you where the

Table 1

Proportion of incident light that is reflected from the flat surface of a lake

	Percent of light reflected	
Incident angle, A	*Vibrating in plane perpendicular to water surface*	*Vibrating parallel to water surface*
0	2.0	2.0
10	1.9	2.1
20	1.7	2.5
30	1.2	3.1
40	0.5	4.3
50	0.0	6.7
60	0.1	11.5
70	4.7	22.0
80	23.9	45.7

angle of incidence is small and how it becomes lighter farther away as the angle becomes larger. Unless you are wearing polarizing glasses, the light reaching your eye comprises both the component vibrating parallel to the reflecting surface and that vibrating perpendicular to it. Compare your observations with Table 1.

Some lakes have something in them in addition to water. Pollution of lakes by man, especially as a result of dumping into them waste products from industrial plants, is deplorable, but it adds interest to the view from the air. Red lakes, yellow lakes and bright green lakes may result from this. The few glaciers that still exist on the North American continent are slowly but steadily grinding rocks together within their icy masses. When the ice melts, this "rock flour" gives the water an apple green color. The fine "rock flour" is white. Why don't glacial lakes appear white? The light entering the lake passes through the water to the reflecting particles and back through the water again, ultimately reaching our eyes. Some of the light is absorbed by the water and this absorption is not the same for all colors (wavelengths) of light. Therefore, light that passes through an appreciable amount of water is colored. The absorption characteristics of water are such that the light that gets through best turns out to be greenish. The white-bottomed waters near coral islands or chalk cliffs look green, as do the waters of white-bottomed swimming pools. It is the combination of reflection of light from white rock particles and selective absorption of light by the water that gives the glacial lakes their milky-green appearance.

Plate IV shows the contrast in appearance from the air between clear water and muddy water. The large light-colored pond is a newly filled pond in which the silt has not yet had time to settle. (The man-made origin of the island is betrayed by its angular shape.) Light scattered from the fine particles of silt makes the new lake look lighter from the air. In time the silt will settle and the large lake will become dark like the small lakes, but the streams will continue to bring silt and their steady contribution will be laid on the floor of the lake, gradually making it shallower. This "silting up" process is happening all the time in all lakes, whether natural or man-made. Sometimes you can see from the air just where the deposits are being laid down (Plate V).

When a reservoir is made by damming a stream, the engineers can predict the rate at which it will become shallower from measurements of the amount of silt carried each year by the streams entering it. In some cases efforts are made to halt some of the silt upstream in small ponds made by damming tributaries to the main stream. The pond shown in Plate

Plate IV. Two small lakes that have been standing for some time and a newly filled lake with fine silt still in suspension in the water

XXXIX is a silt-catching pond on a tributary to Stony Brook, near Princeton, New Jersey. Downstream is Carnegie Lake which will silt up more slowly because of this and other small ponds on its tributaries.

In arid regions some lakes are white around the edges. They are not part of the vast network of connecting lakes and valleys through which, in humid regions, water moves ever onward and downward to the sea. They are in inland drainage basins, receiving water whenever the streams around them flow into them, but giving up water only by evaporation. The arriving streams carry not only solid material in suspension, but also dissolved material, material in solution. If you let a drop of seawater evaporate, the various salts in it come out of solution and stay behind. As the lake loses water by evaporation, the concentration of the water-soluble salts in it

Plate V. Stream distributing silt in a dammed lake

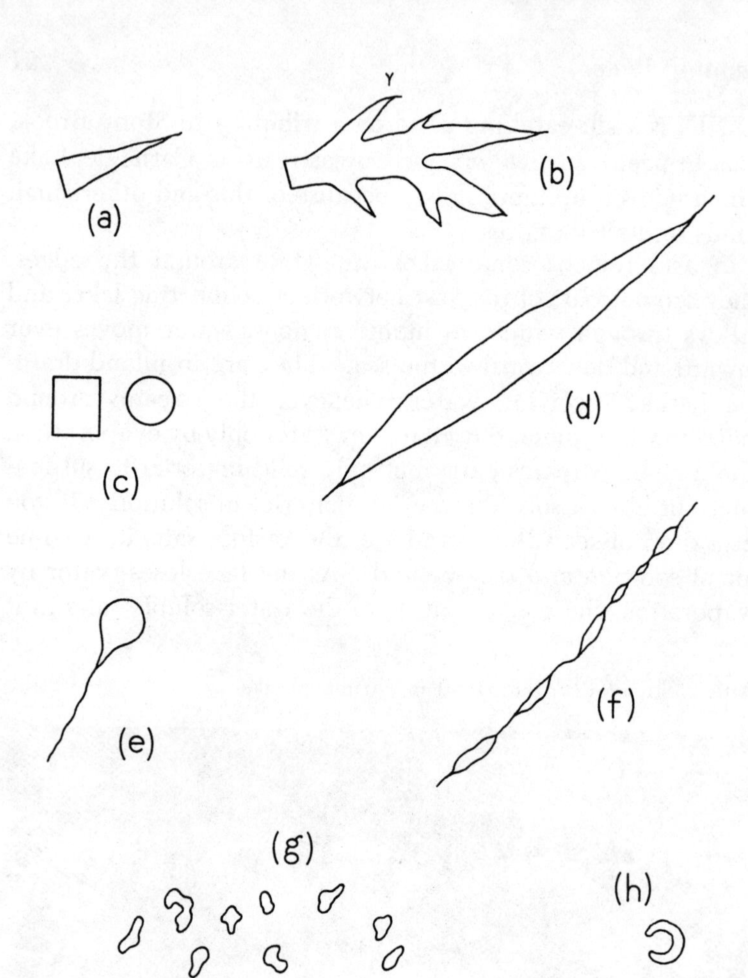

Figure 7. Lake shapes

a,b. Lakes due to damming by man
c. Swimming pools
d. Lake occupying a glacial valley
e. Cirque lake or tarn at the head of a valley formerly occupied by a glacier
f. Lakes along a young river
g. Lakes in irregular glacial deposits
h. An oxbow lake formed by an abandoned river meander

increases. At the edges, where the water spreads thinly as it splashes against the shore, the evaporation is most rapid and a white crust forms.

In the most arid regions such lakes are completely dry during much of the year. From the air, they still look like lakes, except that they happen to contain no water.

What shape is a lake? That, of course, depends on how it got that way. If it has one edge that is a perfectly straight line, man dammed it. Here and almost everywhere in nature straight lines and apparently perfect circles are the work of man, though a lake in the crater of an extinct volcano, like Crater Lake, can be almost perfectly circular. Natural islands never have shapes like that in Plate IV. Figure 7 shows some lake shapes, and the caption identifies them.

Geomorphologists, who, as their name implies, concern themselves with the earth's form, have noticed that lakes tend to become round with time. Wave action shifts sediments along their shores, closing off the mouths of small bays which then get filled in by stream-borne sediments moving toward the lake. Promontories jutting out into the lake are more vulnerable to wave attack than the straight shoreline and so tend to get worn away. Evidence of this rounding process may be detected from the air.*

One place that lakes occur naturally is along the course of a "young" stream, a stream that has not yet worn down the humps and filled in the hollows in its stream bed as it will eventually do. Such lakes are shown in Figure 7f.

The work of ice is responsible for many of our lakes. About 30,000 years ago a vast sheet of ice covered part of the North American continent, flowing southward and westward from

*Some especially clear cases are described by Professor A. N. Strahler of Columbia University in his book, *A Geologist's View of Cape Cod* (New York, Natural History Press, 1966).

eastern Canada and always melting at its southern edge. (See Fig. 8.) Freezing fast to rocks of all sizes, tearing loose pieces of the bedrock, it moved slowly and powerfully across the land, scouring and grooving the surface as it went with the rock fragments frozen in the ice. Sharp hills became rounded by it, narrow valleys broadened. In such a broadened valley the long Lake Cayuga and its parallel sister lakes now lie. (See Fig. 7d.)

As the ice moved, shearing motions within the sheet carried rocks upward into the sheet, even to the surface which in some places was as much as a thousand feet above the base. When this great blanket of ice melted, it left the landscape littered with the debris that it had contained, dropping it here and there in irregular piles and knobs among which rainwater now collects in irregular hollows as though in a bunch of kettles. This kind of topography is called "knob and kettle topography"; its many lakes are small and irregular as in Figure 7g.

In the United States most airlines provide passengers with a fine map of the country. Open it and look at the distribution of lakes. Their abundance north of the Missouri River in the west and the Ohio River in the east is obvious. Compare the airline map with Figure 8. Together they show clearly the work of the ice sheet in disorganizing the orderly drainage of the land.

While the continental ice sheet covered the northeastern United States, some of the mountains of the West had great sluggish rivers of ice, valley glaciers, or "alpine glaciers," deepening and widening the steep valleys on their slopes. Though some of these still remain, some have melted away. At the head of such a valley, where the plucking action by the ice was greatest, you will see a little round lake filling the hollow caused by the quarrying action of the ice. (See Fig. 7e.) Down the valley there may be a succession of several little lakes, dammed by the debris which the ice dropped in the valley as it melted back. Geologists call these Pater Noster lakes, because they are strung along the stream valley like beads on a rosary.

Figure 8. The continental ice sheet

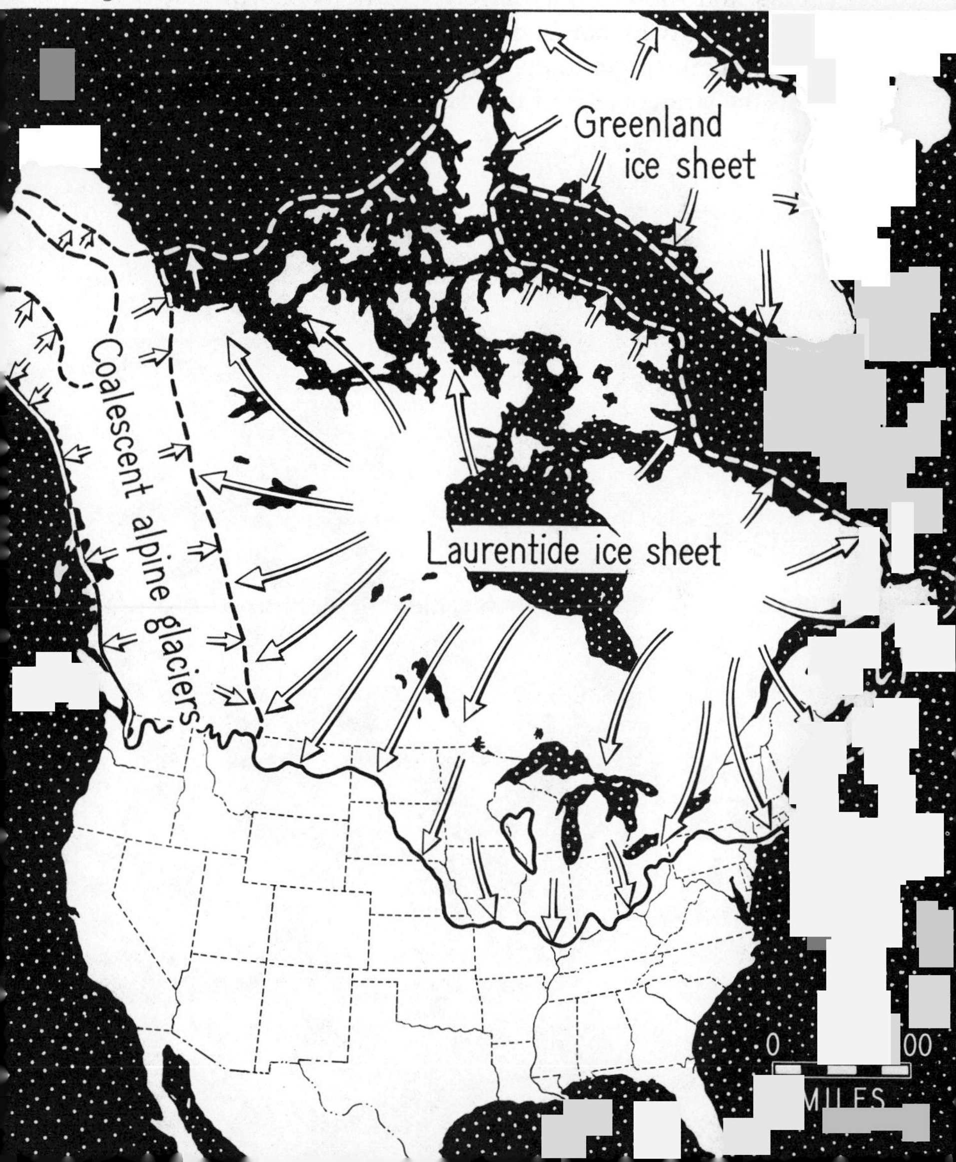

These, like the lakes in Figure 7f, will be drained in time as the water running out of them wears away the obstruction that is retaining them. At the heads of streams new gullies work headward and a system of branching tributaries develops. When a stream is dammed, the lake fills these branching valleys and shows by its fingered shape (Fig. 7b) that it has drowned a river that had tributary streams. Such a shape gives away its origin to the eye of the airplane passenger, even when the straight line of the dam happens to be hidden from view.

3. *Coastlines*

In the sea and large lakes the wave patterns can be especially interesting from the air. In these waves, as in those of the smaller lake, the water moves up and down in circular or slightly elliptical orbits, but not, of course, horizontally forward. If it did, waves would carry ships along with them at their speed of travel, which might be quite unfortunate as they neared the shore, however convenient we might find it in the open sea. Where the water becomes shallow near the land, there is not enough of it to fill out the form of the wave (Fig. 9) and the wave breaks. The water, hurled forward by the energy of the wave motion, moves up onto the beach, carrying with it sand and sometimes inexperienced bathers. Then it flows back into the sea, making a strong "undertow" current that every surf bather is familiar with.

As they approach the beach, waves slow down; their speed is less in shallow water. From the air we can see the white lines of breaking waves crowding closer together near the beach as the faster waves catch up with the slower ones. When waves approach the shore diagonally, with their crests not parallel to the shoreline, the part of the wave reaching the shallow water first slows down first. As a result the whole crest of the wave changes direction (Plate VI). This change in direction of waves due to change in velocity is known as *refraction.* We will encounter it again in the chapter on light. The fact that light exhibits the phenomenon of refraction is consistent with the view that it is wavelike in its behavior.

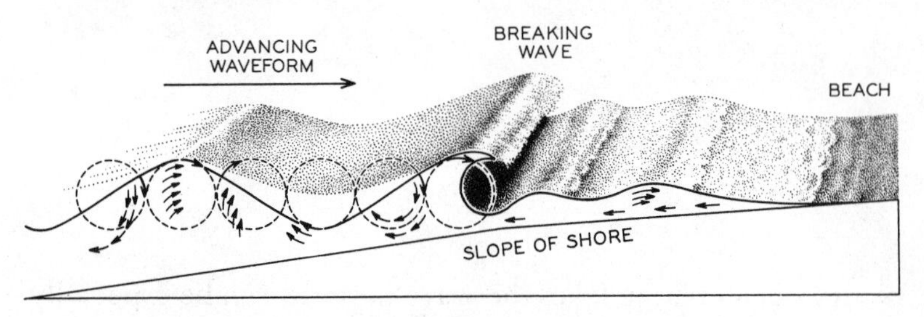

Figure 9. Wave motion and the breaking wave

Even though refraction results in the water waves coming in more nearly parallel to the beach (Plate VI), the parallelism is often not perfect. As the water from the breaking waves moves diagonally against the shore, it shifts sand along the shore. On rocky coasts this shifting sand gets lodged between the headlands in little crescent-shaped beaches, but on sandy coasts it gets dropped wherever currents meet each other, lose speed of travel and consequently lose their power to carry their load. In this way spits and bars and hooks get built up where river water meets the current moving alongshore or where the sea meets itself because of the configuration of the land.

Such structures, built beneath the water by the action of currents due to waves and also, to some extent, to the tides, eventually become exposed at low tide and their sand is piled into dunes by the action of the wind. In this way they grow to be part of the land that is permanently above water. From the air we can see not only those spits, bars, and hooks that have "made it" and become dry land, but also those that are on their way.

Plate VII shows the mouth of the Ausable River at the west shore of Lake Champlain where, century after century, it has been depositing in the lake what it has been removing from Ausable Chasm. It has built a delta, across which the distributaries now make their tortuous ways, still bringing more material which they must drop when they enter the lake, losing

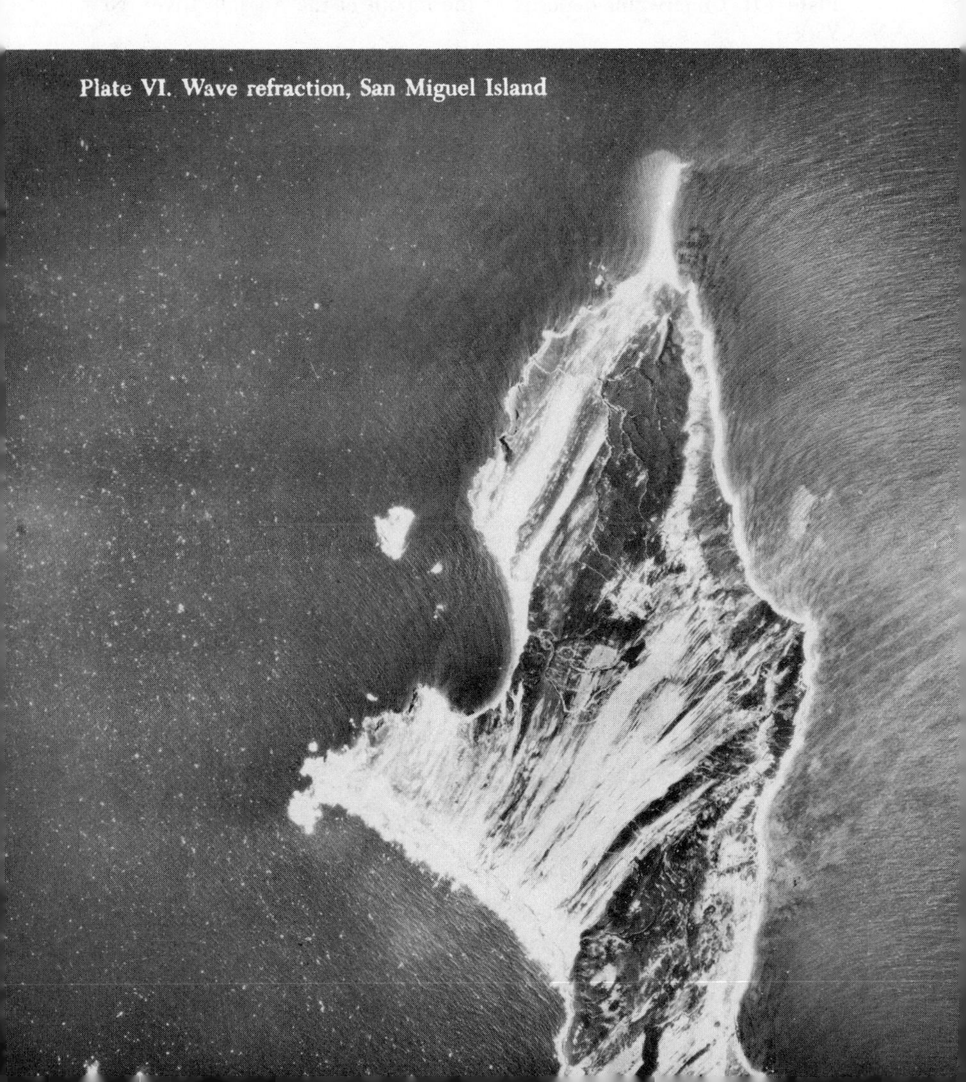

Plate VI. Wave refraction, San Miguel Island

their velocity and their capacity to carry a load. From the air we can see where this is going on under water. Indeed the ridged deposits under water are hard to distinguish, in places, from the whiter sands exposed to the air. The lake-shore currents have swept the deposit from the two distributaries northward and southward along the shore. North of the northern distributary a white spit has been built northward,

Plate VII. Underwater deposits at the mouth of the Ausable River, New York

like a miniature Cape Cod, and beyond it an island with a connection to the spit under water, clearly visible from the air.

As we saw in Chapter 2, our bird's-eye view makes it possible for us to see down into the water in a way we never can from land because of the higher proportion of reflected light at the lower viewing angle. In Plate VIII we are looking directly down on the water in the lower part of the picture and, as in Plate VII, can clearly distinguish the deeper, darker channel from the lighter shallows. The boaters need buoys to mark the channel for them since they are viewing it from a lower angle and these differences are masked by the light reflected from the water. In the upper part of the scene, we too are viewing the water at a lower angle and most of the light that reaches us is reflected from the surface, giving us no information about the depth of the water.

This scene (Plate VIII) and that in Plate IX show the region behind an offshore bar. An offshore bar is a special sort of bar, built in a special way. It is thrown up by large waves breaking farther offshore than the little waves, lifting sand as they scrape bottom (Fig. 9) and piling it landward from where they picked it up until a ridge is built that stands above water at low tide.

Such a bar shows clearly in Plate IX. It has attained some width and is partly covered with a dense growth of stunted trees. The mainland is to the right, out of the picture. Between the mainland and the offshore bar there should be a bay or lagoon. What has happened to it?

The streams entering the bay from the mainland carry, as moving streams always do, a load of sediment. When their velocity is checked as they enter still water, they lose their carrying power and drop their load. If the offshore bar were not there, the ocean currents would sweep it away, redistrib-

uting it along the coast, but the offshore bar creates a still bay and the sediments accumulate behind the bar, filling in the bay.

The flat moth-eaten looking swampland filling most of the scene in Plate IX was formed in this way. It is new land and has not yet had time to establish a drainage pattern. Therefore it is poorly drained, with lakes and little ponds irregularly distributed. What few streams there are cannot cut downward, as most streams do, because cutting power depends on velocity and velocity depends on falling from higher land toward the sea. These poor streams start nearly at sea level. Indeed some of them are just connecting channels from one part of the bay to another and any movement of water through them results from the tides.

Plate VIII. A busy bay

Plate IX. Offshore bar with partly filled lagoon

If this newly formed land can be adequately drained, it may become habitable. In Plate VIII we can see (at the top of the picture) where man has made an effort to hurry this process by digging drainage ditches. The straight lines of these channels, in marked contrast to the naturally meandering ones, indicate that they are the work of man.

Such coastal features as we have been looking down on occur where the land is low and stretches, gently sloping, to the sea and on out under the water. Such a coast has been uplifted from the sea. The sandy regions that are found far inland were formerly beneath the sea. The great slow changes of level that are taking place continuously, without our being conscious of what is going on, come to our attention only through the inescapable evidence that they have occurred.

In some places the level of the land has been lowered relative to sea level. The evidence is clear where we see a valley complete with its tributaries, filled with the sea which makes a bay with tributary arms, a shape often just like that of the dammed lake in Figure 7b. Parts of the land that were formerly hills stand isolated as islands with their feet in the water. From the air we can imagine the water drained away, or the land uplifted so that the river flows again where it used to and the island becomes a hill with its lower slopes dry and grassy.

4. *Running Water*

It is the business of running water to wear away the land. It uses all the particles and pieces that it carries to grind more deeply the valleys in which it flows. Some rivers carry so much that they are colored brown by their muddy load. This may be because their steep descent gives them the necessary velocity to carry it and the area which they drain has much unconsolidated material available for transportation. The Colorado River is brown for these reasons. The Mississippi is not so swift, but it carries to the sea the contributions that many great rivers bring to it. It too is brown with its load.

In some places a hard-working river is a tributary to one carrying less load or a less-loaded tributary joins a muddy main stream. From the air the murky brown waters of one contrast sharply with the clearer waters of the other, and we may sometimes follow the two for some distance downstream in the main river. Plate X shows the confluence of the Tennessee and Ohio Rivers at Paducah, Kentucky, where the clear, dark waters of the Tennessee River join the sediment-filled waters of the Ohio whose particles scatter the light. Figure 10 is a map of the area surrounding Paducah. By

examining what is upstream from their confluence, you will see the reason for the difference in the two rivers. Industrial contamination can also color part of a river for a long way downstream from the source of the contamination.

A river that is in the early stages of wearing down the region through which it flows is called a young river by the geologist, no matter how many thousands of years it has been there. Its banks are steep, its bed not yet smoothly graded (Fig. 11a). It may have rapids in some places, perhaps even falls, and lakes in others. As it deepens its valley, the waters seeping and flowing into it from the sides cause the valley to broaden (Fig. 11b). The river becomes an open-valleyed mature river with smoothly rounded divides separating it from other mature rivers some distance away on either side. In the maturing of a region, streams with more advantages, such as better water supply or steeper courses, eat into and capture the underprivileged streams, so that the region gradually develops fewer, larger streams. As a river nears old age its valley becomes very broad and flat indeed. Its work is just about done and its days of cutting down are over (Fig. 11c).

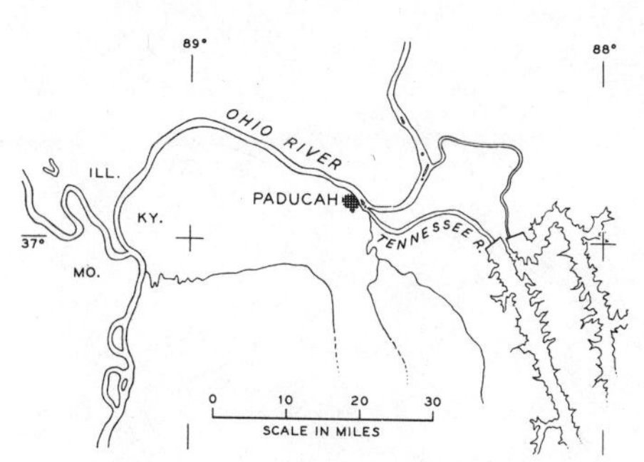

Figure 10. Map of the region near Paducah, Kentucky

Plate X. The confluence of the Tennessee and Ohio Rivers at Paducah, Kentucky

Figure 11.
Young, mature and old streams

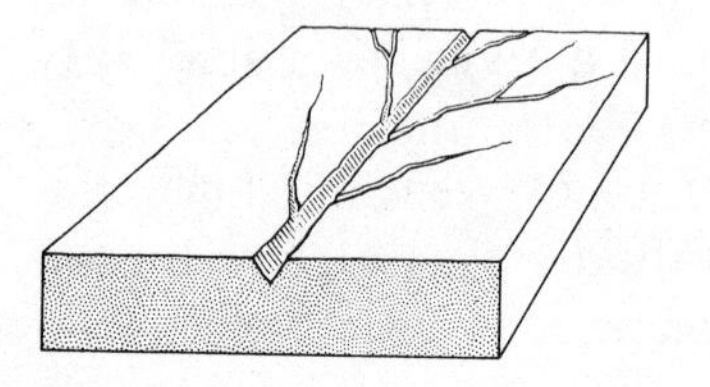

a.

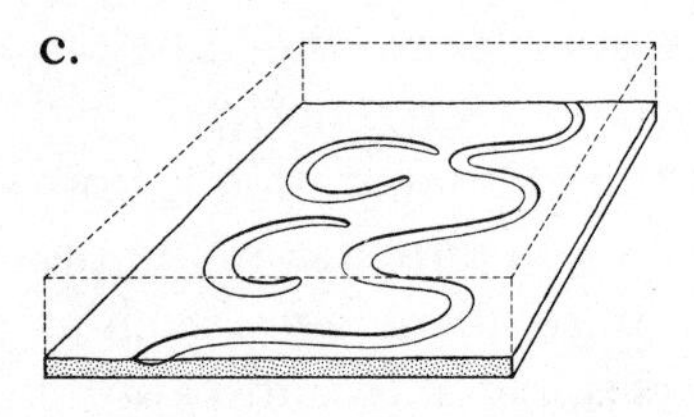

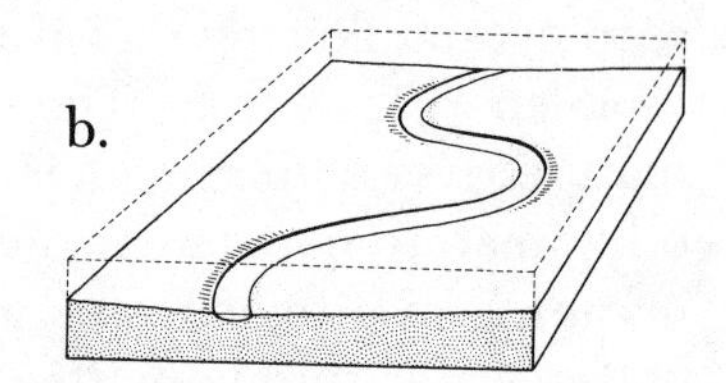

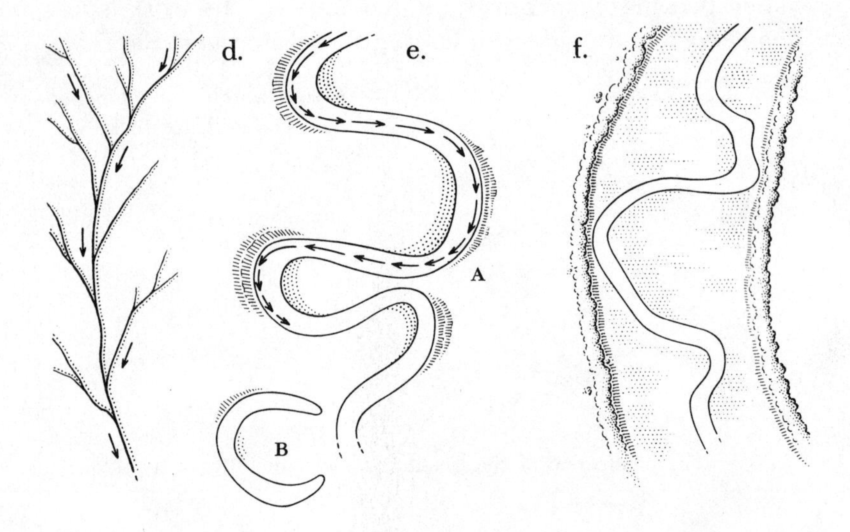

Along with these changes in shape of the cross section of the valley, changes in the drainage *pattern* occur and it is these which we can see so well from the plane. When a stream is young its tributaries join it at sharp angles like the veins in a leaf (Fig. 11d and Plates XI and XII). The rapids and lakes which show that it has not yet smoothed its bed may be conspicuous from the air.

The broad flat divides between young streams are much used for farming. How does such land get drained? Look for the damp spots in the plowed fields (Plate XIII). In some they will be irregularly distributed, but in most they will form a pattern which approaches more or less closely a leaf-vein

Plate XI. Dendritic drainage patterns, young region, Amado, Arizona

or dendritic pattern, depending on how close the field is to an established stream. Far from the stream, the pattern has a salamander-like look. Nearer to the stream a true gully, a baby valley, develops.

In built-up areas the residents sometimes fight with such a nascent stream, filling its little valley with truckloads of soil each year. But the farmer lives too close to nature to fight with her. If he is an old-fashioned farmer, not in touch with

Plate XII. Young streams eating into the high plain south of La Paz, Bolivia

the government's soil conservation agencies, he may just let the gully grow headward into his field and plow around it. Otherwise he will find ways to lead the water across his field so that it benefits his crops as it goes. With terrace farming the water is inhibited from running down the slope by terraces that run across the slope, and there is a shallow, trench-like depression for the water to follow along each terrace. Where the rainfall is great enough so that some runoff from the field

Plate XIII. Pattern of damp areas in a plowed field

must occur, grass-lined open conduits called "grass waterways" are provided and the interlocking roots of the grasses hold the soil against erosion. Watch for these green grass waterways crossing the brown plowed fields below you. Some can be seen in Plate XIV, most conspicuously in the upper left section.

Plate XIV. Farming in a young region in Iowa

The headwaters of any stream are in little gullies that are only streams after a rain. From a plane you can watch for headwaters of streams by following a stream up its valley to the point where you no longer see running water. What lies beyond this? In wooded country you will lose the trail, but in open country you will see how the stream is growing headward (Plate XII) by little gullies that will be small streams in a few decades, rivers in a few centuries.

Beyond these is the divide between the land drained by the stream you have been following and that drained by the next stream. In a young region the divide is a broad flat area. The observer in the plane is in a much better position to recognize it than the observer on the divide itself. You can see where the streams no longer flow toward the east, for example, and start to flow toward the west. How do you tell which way a stream is flowing? When the stream is young, it is easy to tell by the angle at which its tributaries join it (Fig. 11d). In more mature streams the tributary angle may be close to 90° and then this becomes an unreliable indication of direction of flow. If there is an obstruction in the stream — a big rock, a dam, an island — the turbulence it causes will of course appear downstream from the obstruction and so tell us the direction of flow. For the biggest rivers you may cheat by looking at a map.

The mature river has a meandering course (Fig. 11e). How does it get that way? Unlike the tumultuous, rushing young stream, the mature river flows in a more orderly fashion. The motion of the water will be faster in the center where it is not dragging against the sides and bottom of the channel. However, if there is a slight bend in the channel, the faster-moving water, following Newton's first law, will continue in a straight line and run against the outside of the bend. This faster-moving water carries more sand and silt than slower water can carry. Using these cutting tools, it wears away the bank on the outside of the bend, cutting a steep little cliff. Across the river, on the inside of the bend, the water farthest removed from the fast part slows down and must drop its load of sand or silt, building a gently sloping bank. Plate XV shows this process going on in the Missouri River in Montana.

The swifter water on the outside of the bend is sent back toward the center of the river by the curved bank and continues on this course until it meets the opposite bank of the river farther downstream, cutting it back to form the outside of the next meander (Fig. 11e). And so the meandering course

Plate XV. The meandering Missouri River at Fort Benton, Montana

gets started. Once started, it gets more so; the water on the outside always cutting outward, the water on the inside always depositing more of its load to build the bank into the river.

From the air you can often see crescent-shaped sand deposits on the insides of meanders. The steep bank of the outside of the meander may be harder to detect from the air, but in some cases it is marked by shrubs growing on it.

As we have described it, this process of making meanders seems to be a self-intensifying process, a sort of chain reaction in which greater curvature results in more erosion of the bank, which results in greater curvature which results. . . . And so it is. Eventually the curvature becomes so great that the river eats into its own valley, an event that is in imminent danger of occurring at A in Figure 11e. When the break-through is made, the water takes advantage of the shortcut, abandoning its former meander. Such an *abandoned meander* is shown at B in Figure 11e. Another name for it is an *oxbow lake.**

When you fly over a river in the advanced stages of maturity, every step in the remarkable process of meander formation is laid out for you to see. You can see, too, how man adapts to the shape of the river valley. The gently sloping bank of the inside of the meander, covered with the new (geologically speaking) deposits brought by the river, is commonly used for farming. Many examples of this may be seen in western Pennsylvania. A railroad often follows the river be-

*Albert Einstein became interested in meander formation and wrote a paper on it which was published in *Naturwissenschaften* in 1926. He suggested that sediments are shifted from one side of a meander to the other by forces generated in much the same way as those which shift tea leaves to the center of the bottom of the cup when the tea is stirred. For an interpretation of meander formation as the result of a tendency toward a uniform rate of expending energy along a river's length, see "River Meanders" by Leopold and Langbein in the June, 1966, issue of *Scientific American.*

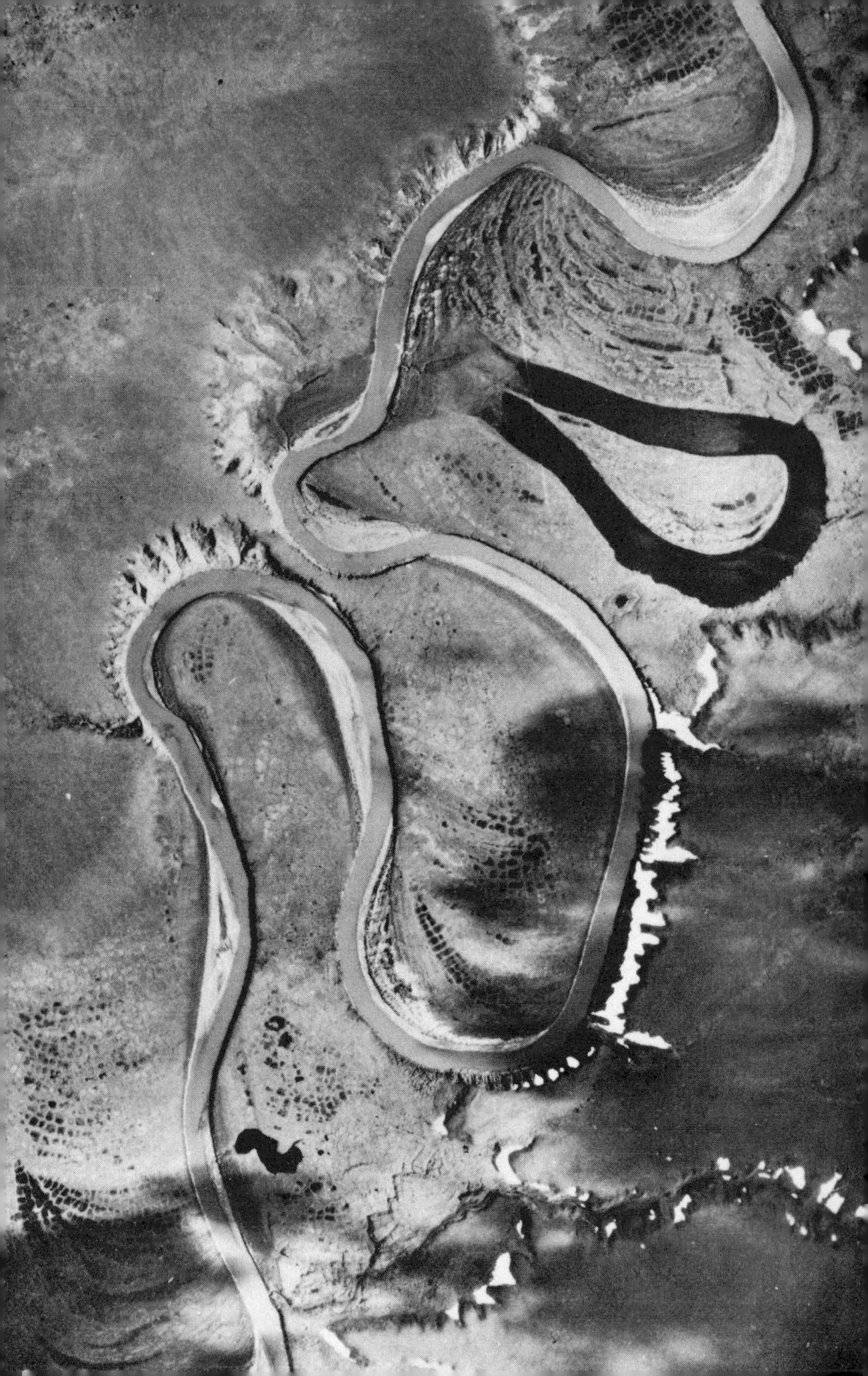

cause the river had already established for it the graded slope that it needs. The problems of a railroad following a meandering river are very amusing. On the outside of the meander it is sorely pinched for space, but to cross to the inside of every bend means building costly bridges. So it does the best it can, sometimes huddling against the bank, sometimes giving up and bridging across to the opposite shore, and always short-cutting across the nearly level land on the inside of the meander.

A mature river shifts its bed repeatedly, probably mostly during floodtime. Far from the present position of the water, you will see evidence of former channels. These may be as distinct as an oxbow lake or as subtle as a curved strip of vegetation that differs slightly in color or texture from that which surrounds it. In Plate XVI the most conspicuous evidence of change of river bed is the dark oxbow lake. Here again the dark waters of the still lake contrast with the lighter waters of the Kogosukruk River. The outside banks of some of the meanders show their steepness by the presence of little gullies. Where these gully scars are far back from the present river they are evidence of an earlier position of the river bank. Curving rows of dark rectangular patches also mark former beds of the river. These are low spots, wetter than the light-colored ground, which owe their geometrical design to the action of ice in the ground during the long Alaskan winter. When the picture was taken the ice had melted, leaving water in the ice-formed depressions.

In some cases the limits of the wanderings of a river are clearly marked by banks that are the boundaries of the valley floor (Fig. 11f). You will detect them by the change in

Plate XVI. Abandoned meander in the valley of the Kogosukruk River, Alaska

character of the vegetation, or by the beginning of vegetation if the valley floor is bare.

Before a river becomes old, it may be rejuvenated. If something happens so that the waters of the river flow more swiftly again and can use their increased burden of sediment to wear the valley floor more deeply, the river will again wear a steep-sided valley, as it did in its youth. This happens whenever the level of the body of water into which the river empties becomes lower relative to the land. Such gradual changes of the relative elevations of land and water are going on all the time without our noticing them. We saw evidence of them in the drowned and uplifted seacoasts of Chapter 3. If the river was already mature enough to have a meandering course before it became rejuvenated, its renewed cutting

Plate XVII. Entrenched meanders of the Little Colorado River near Cameron, Arizona

will take place right in that meandering valley and the meanders will become entrenched as the vigorous river cuts its steep-sided valley (Plate XVII). The tributary streams, less well supplied with water than the main river, may not be able to cut their valleys as rapidly as the river can and may dump their water into the main valley from high on its banks, in thin waterfalls.

There are very few old rivers. In the United States the lower stretches of the Mississippi can, however, qualify for this title. Many miles from its present position you can see evidences of its former channels. Where it enters the Gulf of Mexico, it lays down its burden, building a great delta with the sediment it has carried. Its own waters, blocked by this great mass of debris, find their way to the Gulf as

Plate XVIII. Part of the delta of the Mississippi River

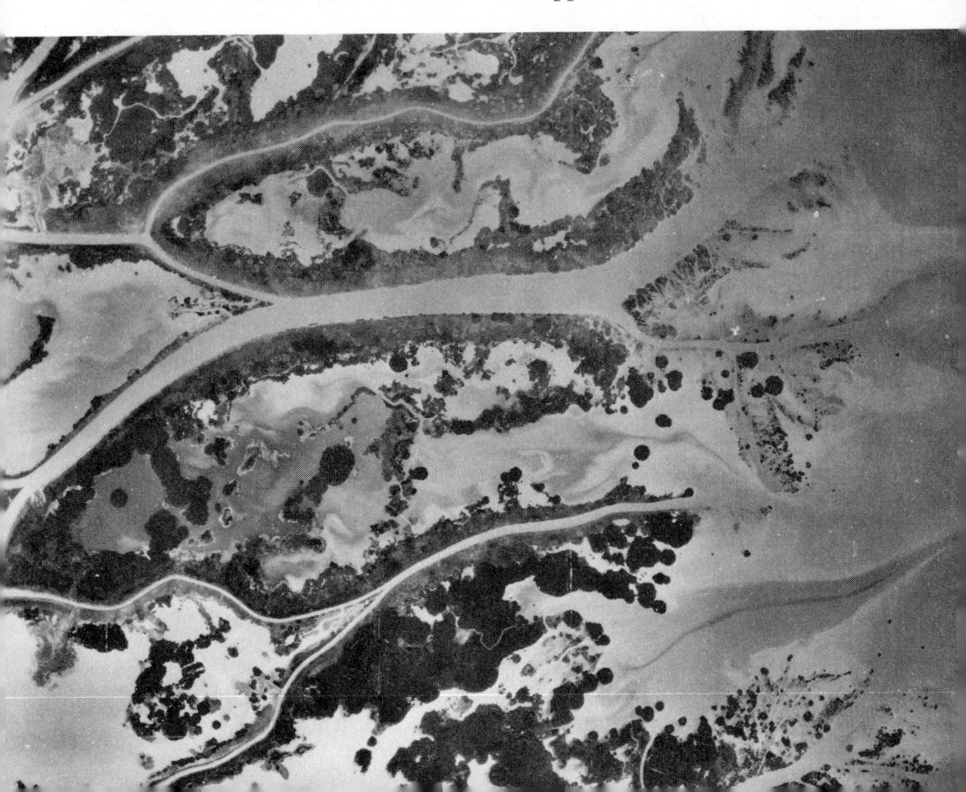

best they can in a system of distributaries, each carrying a little of the load (that keeps inexorably coming from upstream) on farther out into the Gulf. Plate XVIII shows part of the delta of the Mississippi River with its distributaries bordered by the marshland made of their own deposits. In the Gulf of Mexico at the right we can again see the power of the bird's-eye view for detecting underwater topography. Most of the land shown in this picture has been formed since 1900. It is fortunate for Lake Champlain that the Ausable River (Plate VII) builds its delta more slowly.

The old river is not the only river that gets choked by its own deposits. Mature rivers to which tributaries contribute too generously their collections from easily eroded regions may find themselves broken up into a network of small streams which wander here and there over the deposits that the river

Plate XIX. The braided Platte River

could carry only in flood times. It is not hard to see from the air why such a river is called a *braided river.* The Platte River is braided for much of its long length (Plate XIX).

In mountainous arid regions a stream may have a sort of dry delta that is known as an *alluvial fan.* Its fanlike shape is best seen from the air (Plate XX). In arid climates there is much less chemical decomposition of the rock than there is in humid climates. The rock is not transformed into that fine mixture of decomposition products which, with the addition of humus, forms the soil of humid regions. Instead, the disintegration progresses only far enough to result in crumbling of the solid rock into coarse angular fragments.

These crumbled bits of rock accumulate on the hillsides until the rare event of a rainstorm comes to pass. The sudden rivers that result rush down the steep slopes, heavily laden

Plate XX. An alluvial fan

with the coarse debris that was so easily picked up. At the foot of the hills this load is added to the great accumulation of loosely piled deposits from previous flash floods as the river water slackens its speed and settles into the porous dry stuff on the valley floor.

In such regions the bare rock mountains rise from a sea of deposits from their own disintegration (Plate XXI). If some part of a mountain is made of a rock of distinctive color, you may be able to see, at the foot of the mountain, spreading out onto the valley floor, a stain of the same color where the inadequate stream dumped it within sight of its source.

Plate XXI. Mountains standing in a sea of deposits from their own disintegration

Some streams look white from the air. In arid regions this may be due to deposits of various salts as at the edge of a lake. Indeed a large river bed may be dry and caked with such deposits in midsummer. Aside from this, what would make a stream look white? When ocean waves break, the water looks white. Why?

A flat, calm body of water acts as a mirror, reflecting light brightly to the eye of the observer only when it arrives at the appropriate angle, as we have seen in Chapter 2. When a wave breaks, the once continuous body of water breaks up into millions of little droplets, each a sphere or blob of water, separated from the surrounding air by a surface oriented in all directions. These droplets can reflect light from anywhere to anywhere. They scatter any light that comes to them in all directions, one of which happens to be toward you.

One way, then, in which a stream may look white from the air is for it to be a tumbling, frothing mountain stream in which the water is fragmented into millions of droplets, like that of a breaking wave. A group of white flecks on a river may indicate rapids. These are likely to occur where the river crosses those rock layers that are more resistant to erosion (Plate XXXII).

In winter a stream may freeze if it is not flowing too swiftly to give the ice a chance to form without being broken up. If the ice forms in disorderly masses, as it may on a fast-moving stream, it will look white from the air. Winter streams sometimes show white borders where the water that moves less swiftly has frozen.

Larger slower rivers and lakes may become completely ice-covered. When newly formed, such ice is clear and, looking down on it from the plane, we see the dark depths of the lake through it. When changes in temperature cause expansion and contraction of the ice, cracks result. These

cracks appear as white lines across the dark ice which tells us that many surfaces of small ice crystals are scattering the light, as in the case of the water droplets. Snow caught in the crack would look like this, but if the rest of the lake is free of snow, the small crystals may have formed when the crack first opened and the water that had been insulated beneath the ice was exposed to the very cold night air. Under such conditions many very small crystals would form quickly. Probably most white patterns on dark lakes are due to a combination of rapid crystallization of ice and piling up of blown snow.

When snow falls on the ice that covers rivers and lakes, these bodies become white patches on the airplane passenger's landscape and may remain so when the snow on the land has largely disappeared.

A lake that is part white and part dark, like that in Plate XXII, tells a story of a succession of events. White snow lying on old ice covers part of the lake, but new dark ice has formed over a part of the lake that must have been open water during or after the snowfall. The little bays and the water near promontories are shallower and are covered by the old ice and snow.

Ice is a good insulator; it does not conduct heat well. A layer of ice on the surface of a river or lake reduces the rate at which heat is lost to the cold clear sky above and the ice layer grows thicker very slowly as a result. Most substances are denser in the solid state than they are in the liquid state: a given volume of the solid substance weighs more than the same volume of the substance in liquid form. How fortunate it is that the reverse is true of water! If it were not so the ice would sink to the bottom as it formed and entire lakes might freeze solid, exterminating their inhabitants.

Plate XXII. A frozen lake

Where snow stands long on the mountains, lasting through the summer and on into the following winter, it recrystallizes at the bottom into compact masses of ice. As a thick mass of ice accumulates, it begins to flow downward under the influence of gravity. The ice extends into the stream-carved valleys, freezing fast to rocks along their sides. As it continues to flow it plucks away the rocks from the banks of the valley, giving the valley a U-shaped cross section unlike that produced by river work alone. Such a river of ice is a *glacier*. As it grows in length, always fed by the accumulating snow high on the mountain, it becomes dark and dirty with debris called

Plate XXIII. Alaskan glaciers

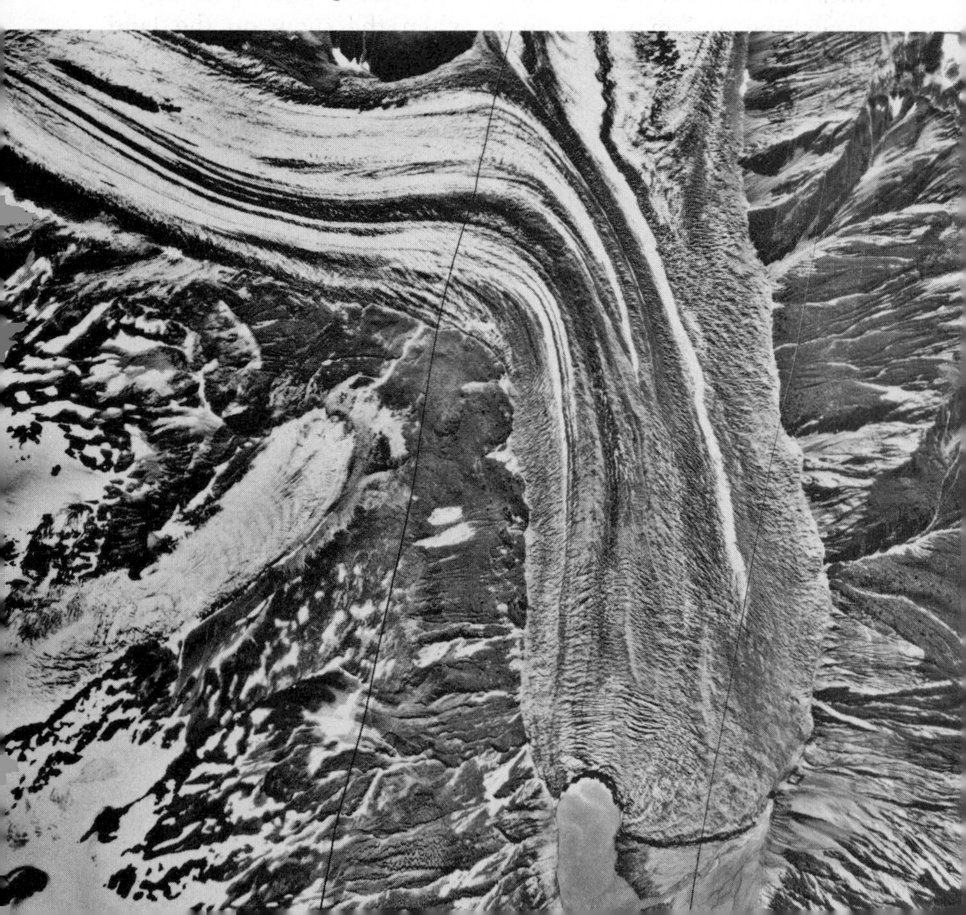

moraine. At its sides, where it is acquiring material from the valley wall, a *lateral moraine* forms. Where two valley glaciers join to form one, their lateral moraines join to form a *medial moraine.* These evidences of glacial action can best be seen from the air. Plate XXIII shows the confluence of two Alaskan glaciers with the formation of a medial moraine. Many little streams tumble down the steep banks to bring water-borne sediments to the edge of the glacier. At the right a little river runs along the edge of the ice, dumping its load in a fan-shaped deposit where the speed of the water is checked at the end of the ice.

The smoothly rounded end of the glacier has a bite taken out of it where masses of the ice have broken off and floated away as icebergs in the lake. The sharp cliff left by the break casts a black shadow on the water.

To the left of the main glacier is the white tongue of a smaller glacier flowing down from the high peaks farther to the left. Note the faint curved bands of moraine on its surface. These show that the ice in the center moves ahead faster than the ice that drags against the sides of the valley. Such motion may be less than a foot a year near the sides and perhaps two or three feet a year in the center. This does not mean that the tip of the tongue is approaching the main glacier at this rate, because it may be melting away as fast as or faster than it is proceeding forward. Frothing white rushing streams run out from the tip of the white tongue of ice and flow down the steep slope to the main glacier. These streams are too small to carry the largest rocks that the ice has carried in its frozen solid mass so these accumulate at the end of the melting ice, forming a *terminal moraine* which can be seen at the tip of the small white glacier.

If you stood on the surface of the glacier you could break off pieces of ice from it with a hammer. They would be as brittle as the ice cubes in your refrigerator. From the air we see abundant evidence that this brittle substance has been plastically deformed, changed in shape, as it moved down the valley over a long period of time. In Chapter 10 we will see that the hard rocks of which the earth's crust is made have also been deformed without breaking. But it takes time.

5. *Light*

JUST BEFORE YOU TAKE OFF, a glance at the top surfaces of the plane's wings would convince you that they were silver-colored, brightly metallic or even whitish in appearance. As you climb to 30,000 feet or more, look again. You will find that they have become a dull, sooty dark gray. How does this happen?

The light that was reflected to your eyes from the wings when you were on the ground came through the lower atmosphere. The lower atmosphere is full of tiny particles of dust and sometimes water which scatter the light just as the water particles in a breaking wave do. It is the scattered white light from this lower-atmosphere haze that is reflected to your eyes by the wings when the plane is on the ground. When the plane has climbed above this haze, the top surfaces of the wings "see" only the clear, clean very dark blue sky (look *up* at it from the plane!) and they look dark. Why don't they look dark blue, like the sky? They would if they reflected as a mirror reflects, showing you the sky above. Apparently the fact that the surface is not mirror-smooth has to account for the color not being blue. The fine-textured roughness of the wing surface must scatter the blue light in all directions and only a small portion of it reaches your eye. This is an inadequate hypothesis to explain the total lack of blue color, but it is a step in the right direction.

Why is the sky blue? It is an old question and a favorite in school science classes although it is not an easy question to answer. Old textbooks tell you that it is blue because the fine particles of dust and moisture in the air scatter the blue (shorter wavelength) light more effectively than they scatter the redder (longer wavelength) light, but physicists have found that light-scattering by the air molecules themselves accounts for the hue and intensity of the blue of the sky, whereas scattering by particles larger than molecules detracts from the blue color and gives the sky a milky appearance.

Compare the color of the land directly below you with that of the farthest hills. The preferential scattering of blue light by the air makes the distant mountains look quite blue because the light reaching our eyes comes not only from the mountains, but also from the blue-scattering air that lies between. Light falling from above onto the air contributes scattered blue light to all the rays that reach you from distant objects.

When you fly at night, especially if it is a hazy night, compare the color of the lights directly below you with that of the farthest lights that you can see. The sprawling bright spider webs of lights below you are yellow-white, while the flat little pancakes of lighted towns in the distance are reddish. Since red and blue are at opposite ends of the visible spectrum, this daytime-nighttime comparison seems at first to constitute a strange and contradictory pair of observations. The clue to the explanation lies in a simple experiment which is so rewarding that everyone should try it.

For this experiment you need a drinking glass full of water to which a few drops of milk have been added. First look at it with the light coming from the side. With your side toward the window, hold the glass straight in front of you, about at eye level. The milky water looks bluish. Next turn and look toward the light through the milky water. It changes to pale orange color!

In both the milky water and the atmosphere the very small drifting particles scatter the short-wavelength blue light more effectively than they do light of longer wavelengths. The distant lights on the ground send out the same light as those directly below you. The wavelengths of the rays emitted constitute a continuous range from the short-wavelength violet and blue through green and yellow to the longer-wavelength orange and red. But all the way from the distant light-source to you the shorter-wavelength light is being scattered in all directions more than the longer-wavelength light is, so that there is less blue color and more red color in the beam by the time it reaches you. Of course this is also happening to the near lights, but their light travels a shorter distance and loses less blue.

If you are flying in the bright white sunlight above a bluish haze that obscures the land below you, you may suddenly see the reflection of the sun in a body of water and find that its color is the rich red gold of the setting sun. The reflected light has traveled down through the blue-scattering haze and back through the blue-scattering haze again to your eye. Thus impoverished in the bluer light, the light that reaches your eye is rich in red and gold. To duplicate this effect with the milky water, put a shiny spoon in the glass. As you look at the blue milky water, lighted from the side, the bowl of the spoon will shine with a ruddy light.

In explaining the change in color of the wings, we said that the scattered light from the atmosphere near the ground was white. Now we are saying that most of the scattered light is blue. The color of scattered light depends on the size of the particles that scatter it. Smoke curling up from the lighted tip of a cigarette scatters blue light predominantly because of its very small particles. Cigarette smoke that has been held in the mouth looks yellowish white when it is exhaled. The particles have become coated with water molecules and are too large to scatter mostly blue light.

The whitish scattered light that is nearly always part of the ground illumination comes from tiny water droplets and dust particles. Even the bluest sky looks whitish near the horizon because there the light travels to your eye through more of the dust-filled lower atmosphere.

In Chapter 2, in discussing the polarization of light reflected by water, we came upon the observation that light from the sky is polarized. Nowhere can we see this so well as from a high-flying plane, when we are above the dust and haze which scatter the light repeatedly, destroying its polarization. Some experimenting with polarizing glasses will convince you that the polarization of the sky's light is greatest when you are looking in a direction 90° from the sun's rays (Fig. 12). The direction of vibration (electric vector) of the light is as it would be if the light were being reflected from many little reflecting planes, that is, it is perpendicular to the plane in which the ray travels from the sun to the sky to your eye (perpendicular to the paper in Figure 12). Test this yourself.

When you look at the sky with your polarizing glasses in the proper position to let through the minimum amount of the sky's polarized light, the clouds stand out in bold contrast. Clearly the light from them is not polarized. What about the finer blue-scattering haze? Does it behave more as the blue sky does or more as the clouds do?

The light scattered by the haze is indeed polarized, as you can verify by looking through haze at the ground, using your polarizing glasses oriented in various positions. This is why photographers use a properly oriented polarizing filter to "cut the haze" for distant shots. For cloud photography, such a filter, oriented so as to give the darkest sky, provides maximum contrast between the clouds and the sky.

This is probably a good place to discuss some of the problems that arise for photographers who wish to take pictures from the airplane. The first problem is the choice of a seat. The sunny side offers an opportunity to photograph bright reflections from water, but the sun highlights all the scratches and dirt on the windows and may cause a bright reflection of your camera in the window as well. On the side away from the sun, the window will cause you much less trouble

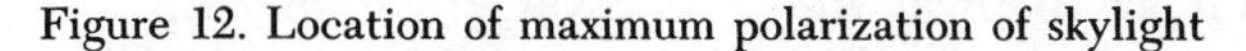

Figure 12. Location of maximum polarization of skylight

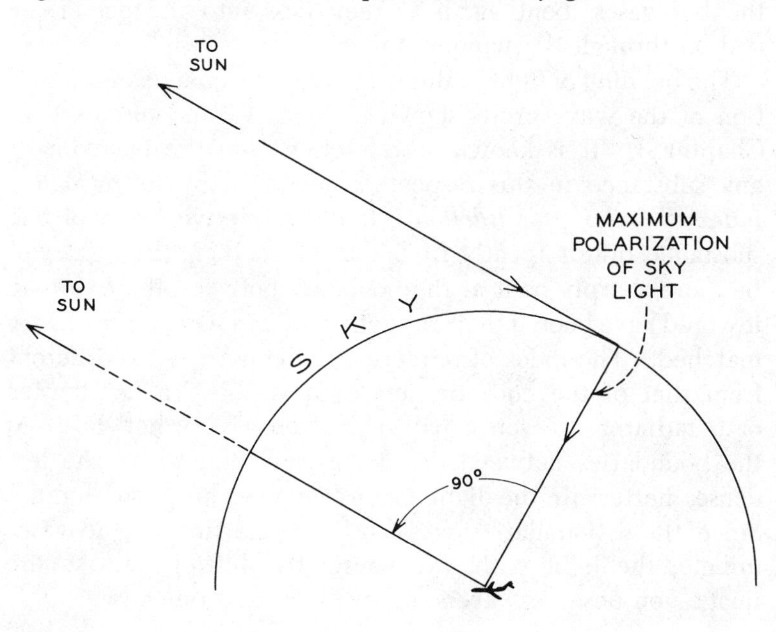

unless it is very dirty. (Some airlines keep their windows clean and some do not.) Furthermore, the shady side gives you an opportunity to photograph the splendid shadow and glory effects discussed later in this chapter.

A seat ahead of the wing gives you the best view, but if you do not travel first class, as I do not, this may be difficult to manage. If the wings are swept sharply back, a seat very near the front edge of the wing is entirely satisfactory, but it is not good in a propeller plane unless you want to take pictures through the propeller.

If you sit just behind the wing in a plane in which the motors are wing-mounted, you may be looking at the view through the mass of gases that pour out from the exhaust pipes. That view will be blurred. The light rays from the distant objects are bent as they pass from the cooler air into the hot gases, bent again as they pass into the cooler air and on through the window to you.

The bending of light in this way is like the change in orientation of the wave crests shown in Plate VI and discussed in Chapter 3. It is known as refraction, and the behavior of any substance in this respect is measured by its *refractive index* or *index of refraction.* If the refractive index of one substance differs greatly from that of another, the light will be more sharply bent at the boundary between the two than it would have been if their refractive indices were more closely matched. The index of refraction of the hot gases is different from that of the cool air, just as it is above a hot toaster or a radiator or even a paved road on a very hot day. At the boundaries between the denser, cooler air and the less dense, hotter air the light beams change direction slightly. Since these boundaries are constantly shifting as the gases mingle, the light paths are constantly shifting, and so the image you observe wavers and wobbles and blurs.

In a plane with wing-mounted motors, therefore, the photographer would do well to sit as far back from the wings as possible when he cannot sit ahead of them.

If, on arriving at your carefully chosen seat, you find that the inside of the outer windowpane is clouded with condensed moisture, do not despair. It will probably go away when you are in the air. The reason for this is discussed in Chapter 6.

You will get the best ground photographs while you are still less than a mile up. Except on unusually clear days or in desert regions, so much light is scattered by the haze in the lower atmosphere that hazy photographs result. What about the use of a polarizing filter to "cut the haze" as just described? In most windows, the strain or structure of the plastic, discussed in Chapter 2, is such that the window will contribute color bands to any picture taken with a polarizing filter. If you like the added color, then you have no problem.

There is yet another way to reduce the haze problem. A filter that has a yellowish tint will selectively reduce the amount of blue light passing through it. Such filters can also be made so as to absorb the scattered ultraviolet light as well which, though invisible to the eye, is recorded by the film. A number of filters of this type are made. For color photography from the air, Eastman Kodak recommends Skylight Filter, Type 1A; for black and white pictures there is a series of three filters, given here in order of increasing absorption of a short-wavelength radiation (blue-violet end of the spectrum): K2, G, A.

When conditions are good for cloud photography you may often find that the light intensity is very high. You must believe the reading on your exposure meter. The results can be spectacular. However, when photographing the ground, give more exposure than your sky-influenced meter indicates.

The blurring of images by the exhaust gases was mentioned as a nuisance to photographers. The shifting of light paths due to mingling gases of different densities is also responsible for the twinkling of the stars and of distant lights. From the ground, on a still, clear night, the stars straight above you may hardly twinkle at all. The stars near the horizon, however, are nearly always twinkling because the light from them comes to us slantwise through the atmosphere, passing through so much air that it is sure to encounter unevenly heated regions with shifting boundaries between warmer and cooler air.

From our airborne observation post, flying at night, we are far from being still observers. This alone may result in such variability in the light path to us from some ground light that it twinkles, but, as in the case of the stars, the lights near the horizon are likely to twinkle more.

Above us we see the stars through the relatively homogeneous thin air of the stratosphere. Do they still twinkle? Perhaps the old nursery rhyme needs to be brought up to date.

> Steady, steady, little star,
> What a constant lamp you are!
> Far below, each man-made light
> Twinkles up at me, in flight.

Just after takeoff, the shadow of the plane shows its shape clearly. As we climb higher the shape of the shadow changes and finally, just before we lose it altogether, it is circular! This is a phenomenon familiar to you in another form. When the sun shines through the leaves of a tree the spots of light that dapple the ground are all elliptical. The beams, whose cross sections are circular, meet the ground at an angle and sprawl into ellipses of light. The openings in the tree through which these beams of light come are irregular in shape, but small. They act as pinholes because they are small compared to their distance from the ground and each produces an image of the sun as any pinhole will do. You can reproduce the effect by cutting a small irregular hole in a large piece of cardboard and, using a round bare frosted light bulb as a light source, observe the image cast onto white paper held at various distances. The results are even better if you use the sun itself. You will see its circular image whenever you have a small irregular opening in a large area that is obstructing the sun's light.

Your plane is just the same, but opposite: a small irregular obstruction in a large space that is not obstructing the sun's light. From each point on the circular sun, light casts a shadow of the airplane. A light beam from one rim of the sun makes an angle of about half a degree with a light beam from the other rim of the sun. Therefore the shadow cast by light from the left rim does not coincide exactly with the shadow cast by light from the right rim. When the airplane is close to its shadow — that is, close to the ground — the half-degree difference in the direction of the beams results in an imperceptible displacement of the shadow. However, when the plane is far from the ground, shadows cast by rays half a degree apart will no longer approximately coincide. There will not be separate shadows, because light is not just coming from the two points on opposite sides of the sun, but from an infinite number of points over the whole round sun. The spot is made up of an infinite number of shadows, the outermost ones, all around the rim, being made by the outermost sources of light around the rim of the sun. Only where all the shadows overlap will the shadow be dark. This composite shadow is circular because the sun is circular. If you are ever so fortunate as to be in an airplane that is taking off during a solar eclipse, watch for a circular shadow with a bite taken out of it!

As you watch the dwindling shadow of the plane you may find that it is surrounded by a brighter region and eventually the shadow may give place to a small bright spot! This is best seen when the shadow of the plane passes over grassy fields or over clouds and is probably the Heiligenschein described by Minnaert* as visible around the shadow of a balloon. He shows a photograph of it around the shadow of the head of the photographer cast by the early morning sun onto dewy grass. According to Minnaert, "the sun illuminates most of the stalks [or droplets] through the spaces between the foremost rows; anyone looking more or less in the direction of the sun's rays will see all the small illuminated surfaces; if he looks more sideways he will see many blades [or droplets] in the shade, the average brightness, therefore, being less."

When you fly above a cloud layer, even a thin one, the occupation of watching the plane's shadow can be filled with exciting variations. As the character and distance of the clouds below you change, the bright spot turns to a shadow and back to a bright spot again. Suddenly a very near cloud shows a large shadow of the plane, with every detail of its outline sharply visible. Then the reflecting surface falls away and perhaps neither shadow nor bright spot can be seen. The most spectacular event in the course of this occupation is the appearance of a *glory*. A glory is a rainbow-like ring completely encircling the shadow of the plane, or the place where the shadow would be if there were one. Having learned how to look for one, I now see it on nearly every plane trip when I sit on the shady side of the plane. You first notice

* M. Minnaert, *The Nature of Light and Colour in the Open Air.* (New York, Dover Publications, Inc., 1954).

a pinkish tinge to some of the clouds, especially those clouds drifting in thin veils above a lower solid-looking layer. Then you become aware that the color is arc-shaped and finally you see the whole rose-colored circle, perhaps with a tinge of blue on the inside.

Under very favorable conditions you will see a double glory! A second colored ring partly or completely surrounds the first. If you are familiar with double rainbows, you will notice an important difference between a double glory and a double rainbow. In a double rainbow, the sequence of colors in the outer bow is the reverse of that in the inner bow so that the two red bands face each other, one on the outer side of the inner bow, the other on the inner side of the outer bow. In the double glory, however, the sequence in both rings is the same, with red always on the outside.

If the cloud layer producing the glory is near enough, you will see the shadow of the plane in the center of the glory, more or less definite in shape, according to your distance from the clouds. If you are fortunate enough to be watching a glory as the plane starts its descent, you will see the shadow of the plane get larger and larger until it fills the whole ring of the glory. What happens then? It keeps on growing and you see that the center of the glory is determined, not by the plane, but by you. Wherever you are sitting in the plane, the glory will be centered around that point in the shadow that marks your position.

Color Plate I, on the inside front cover, is a photograph of a glory taken from a seat just behind the wing. The cloud causing the glory is close and the shadow of the plane looms large. The center of the glory shows my location as I took the picture. Where the shadow of the wing joins the shadow of the body, part of the central bright spot shows where the sun's rays are reflected straight back to the camera.

Long before man flew airplanes he saw his shadow on the mountain mists when he climbed to high places. In the late eighteenth century, climbers in the Harz Mountains noticed this especially when they stood on top of the highest mountain in northern Germany, the Brocken. The dark figure outlined on the mountain mist may have had a glory around its head, or at least a Heiligenschein, and seemed like a ghost to the mountaineer who saw it. Because of this it was named the "Specter of the Brocken."

The center of the glory is the point where a line drawn from the sun through your head meets the cloud. The rays of colored light from the glory circle to your eye form a cone of rays, with your eye at the apex of the cone. All this is also true of a rainbow, but, as we shall see later, the angle of the cone of the rainbow is always the same (Figs. 13 and 14), whereas that of the glory is not. The angle of the glory varies inversely with the size of the droplets causing it. If they are large, the glory circle is small (that is, the angle of the cone is small). Indeed, you may sometimes see one part of a glory bulge or shrink a bit as a cloud with finer or coarser droplets takes part in its formation. The glory will be sharpest when the droplets are uniform in size. The sequence of colors and the dependence on droplet size that characterize the glory show that it is a diffraction phenomenon, resulting from constructive and destructive interference among the light waves.*

Watching shadow and glory phenomena from a plane is an occupation filled with suspense and sudden rewards.

*Early attempts to understand the glory failed because it was thought that its angular size was related to the droplet size by the simple diffraction equation. Research has shown that this is not so. For an up-to-date discussion of this, see *Introduction to Physical Meteorology* by Hans Neuberger, published in 1965 by the Continuing Education Division, The Pennsylvania State University Press, University Park, Pennsylvania.

As more people find out about this, the window seats on the shady side of the plane will become increasingly hard to get.

Some optical phenomena require such special conditions for their observation that they are only rarely seen. "But," as Shakespeare had Prince Hal say of holidays in *King Henry IV*, "when they seldom come, they wish'd for come, and nothing pleaseth but rare accidents."

Once I saw a very clear effect at the leading curved edge of the wing due to the difference in index of refraction between a layer of air close to the wing and the air farther out. A straight road below me appeared to bend sharply an inch or two in front of the wing. As we passed over the various roads and houses every image would have a sharp bend in it when I saw it through that region of air just ahead of the wing. My current hypothesis is that I was observing the effect of the boundary between the thinner air passing close over the curved surface of the wing (Fig. 1) and the denser air farther from the wing.

A friend of mine has described to me a phenomenon which I have never seen, but which he assures me can often be seen, once you know how to look for it. The fastest passenger airplanes travel at speeds close to, but just less than, the speed of sound. This means that the speed of air moving past the plane is close to that of sound in that air at that temperature. Over the top of the wing, however, the air is moving a little faster, relative to the plane (Fig. 1) and can attain the speed of sound. To put it another way, part of the surface of the wing is moving at the speed of sound relative to the immediately neighboring air.

According to my informer, who is a competent scientist and careful observer, this results in a density discontinuity and therefore an index-of-refraction discontinuity which can be seen in the following manner. If the sun is shining onto the wing from behind you, the deflection of its rays because of the discontinuity in index of refraction produces a clearly defined line of shadow along the wing, parallel to its long edges. Because of the nature of the air flow over the wing, this line of shadow commonly branches near the outer tip of the wing. If the plane tips or changes its orientation slightly the pattern may break up and disappear, only to form again when the plane reestablishes its course and speed.

A rainbow seen from the ground as we stand with our backs to the sun (or to the moon) is a splendid sight, but it is incomplete. Only when we look down on it can we see the whole rainbow. Each ray of light that comes to our eyes from the rainbow has entered a drop of water in the air, been reflected from its far side and come back out again. It is bent or refracted on entering and leaving the drop and the different colors of light are refracted differently. In other words, the index of refraction of water varies with the wavelengths of the refracted light, as does that of many other substances.

The geometry is such that for each ray of refracted light there is an angle at which its intensity is a maximum. The analysis is not simple since the result cannot be achieved by considering the path of a single ray, but the effect is understood. For red light, on the outside of the bow, the angle of maximum intensity is about 42°. The "returning" rays make an angle of 42° with the entering rays from the sun. In Figure 13 the straight dashed line represents that particular ray from the sun that would go right through the man's head if he didn't stop it, and its extension in a straight line on the other side of him to the *antisolar point,* the point opposite the sun, the "shadow point" of the observer's head. The red rays of the rainbow that come to the man's eye make an angle of 42° with this line because it is parallel with the ray from the sun that entered the drop of water. Everywhere at 42° to this line, the man sees red rays from water drops. Rays of other colors, at smaller angles to this line, lie inside the red. These rays come from other drops. It is as though the rays followed the surface of a cone whose apex was at the man's eye and whose axis was the line connecting the sun and the antisolar point. As the sun falls lower in the sky, the rainbow rises higher, but only part of the cone will be above ground. He can never see the whole rainbow unless he is above it.

Figure 14 shows the same 42° cone along which the rainbow rays travel to the observer's eye, but here the observer has

Figure 13. The rainbow, seen from the ground

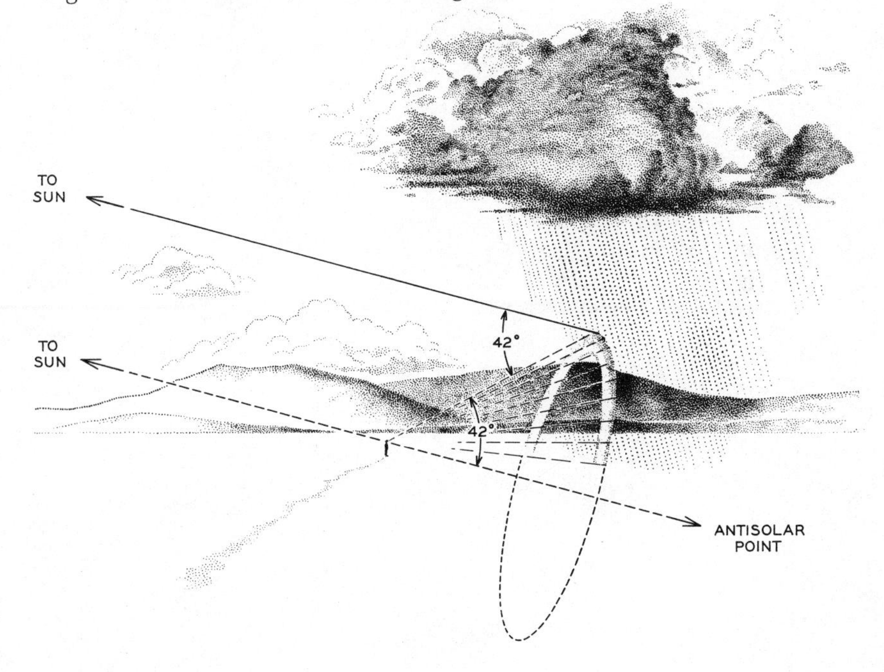

sunbeams passing below him as well as above him. He can see the lower part of the rainbow as well as the upper part; for him the rainbow is a full circle.

Since every ray of the rainbow comes to the eye of the observer by reflection from the back surface of a water drop, what would you expect about its state of polarization? If you keep a piece of polarizing film (or polarizing glasses) with you during the rainbow season you can determine for yourself whether your expectation is fulfilled. The results of this experiment may surprise you.

Figure 14. The rainbow, seen from the air

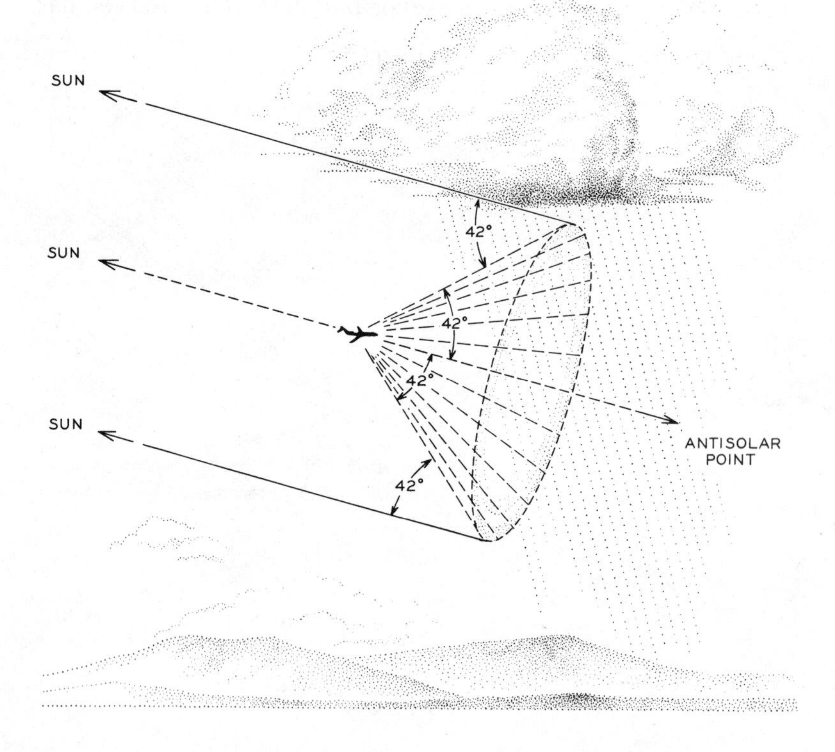

Rings of light around the sun or moon are called *halos.* When high, wispy clouds come between you and the sun or moon you may see a halo with a radius of 22°. These clouds are made of very small ice crystals drifting in the upper atmosphere and the halo results when rays of light from the sun or moon pass through the crystals and are bent because the index of refraction of the crystal is greater than that of air. The ice crystal is built up by the orderly arrangement of water molecules, layer upon layer, closely bonded together to form a solid. Because of this orderliness it has a definite shape and is bounded by plane surfaces meeting at definite angles determined by the arrangement of the water molecules of which it is composed. The light entering each crystal is therefore refracted in the same way, the refracted beam deviating from its original direction by about 22°. Since the index of refraction of ice varies with the wavelengths of the refracted light, the colors will be spread. However, the shapes, sizes and orientations of the crystals are not so uniform as those of the raindrops, so the colors overlap and are less distinguishable than in the rainbow. Red light is bent least and the inner border of a halo is therefore reddish.

Figure 15 shows how both the ground observer and the airborne observer (if he is flying low enough to have any clouds above him) will see this halo as the light rays come to them along the surface of a cone with an axial angle of about 22°. In addition, the airborne observer has the chance of seeing the antisolar halo formed by refraction and reflection of light from ice crystals on the far side of the plane from the sun. The geometry is such that this halo occurs at about 38°. It is a weak halo, seldom seen.

On the back cover of this book is an instrument with which you can measure the angular magnitudes of any light rings you may see from the plane. The best procedure would be to measure the angular magnitude from side to opposite side of the whole ring (44° in the case of the 22° halo, for example) and then divide by two to get the cone angle. This is better than measuring the cone angle directly for two reasons. First, it may be hard to determine directly where the center is from which you would have to measure. Second, if the measurement of double the cone angle is uncertain by plus or minus four degrees (±4°), the measurement of the cone angle, which is what you want to know, is uncertain by only ±2°. In other words, for a given absolute error in measurement, your percentage error will be smaller if you measure a larger quantity of the thing measured.

Figure 15. Ice-crystal solar halo at 22 degrees and antisolar halo at 38 degrees

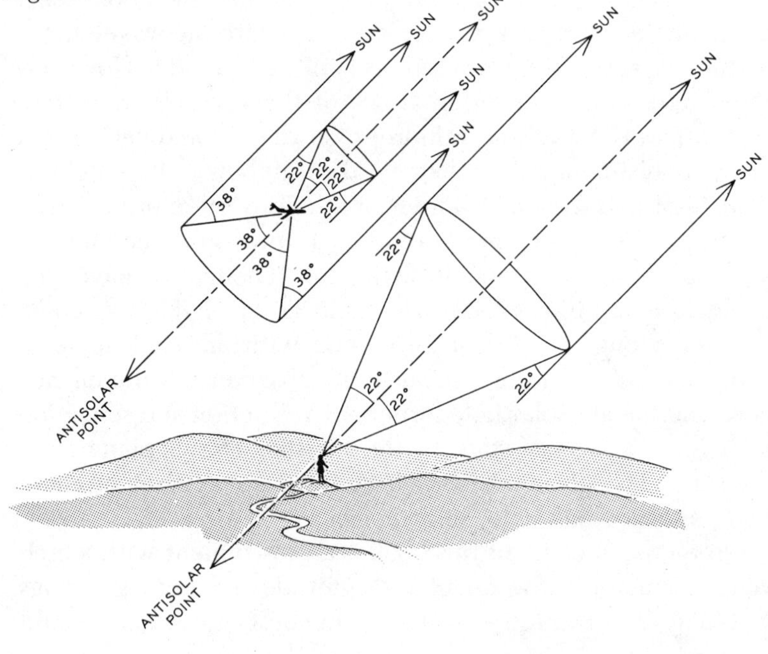

An instrument for measuring angles is called a *goniometer*. The technique for using this instrument is illustrated in Figure 16. Sight one side of the ring along the spine (zero edge) of the book with your eye right at the corner. At the same time sight along a pencil laid on the book with one end at your eye and the other end pointing to the opposite side of the ring. If the pencil doesn't slip between the time you sight along it and the time you read the angle from the instrument, then your angle reading will be a valid one.

A string attached to the corner of the book from which the goniometer lines radiate would enable you to make better measurements than you can make with the pencil. Smooth,

Figure 16. Technique for using the bookback goniometer

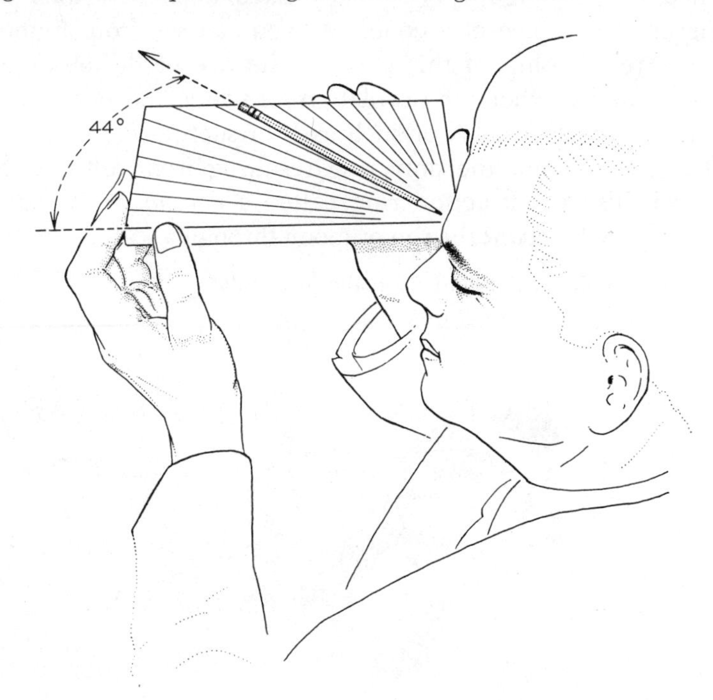

straight cord, such as nylon fishline, would serve best. It could be glued on the inside back cover and allowed to hang out at the top. In between measurements it could be used as a bookmark, like the red ribbon that was a part of my grandmother's Bible.

An interesting property of all these ring phenomena is that they are always circular, even though the cloud layer where they originate is a horizontal layer and the sun's rays are reaching it obliquely. One might suppose that a glory formed by obliquely incident rays on a horizontal cloud layer would appear elliptical. The diagram in Figure 17 shows why it does not. The rays coming to your eye travel a path that is at a particular angle to the sun-antisun line (the dashed line in Figure 17). They have been described as lying on the surface of a cone. As you can see from Figure 17, the relationship of this cone to the eye of the observer is the same, whether the droplets from which the rays come lie in a plane such as A or such as B or such as C.

The glory is not the only ring resulting from diffraction. There is also a diffraction ring called a *corona,* to be seen when you look *toward* the sun or moon through a water-droplet

Figure 17. Rainbows and glories are always circular

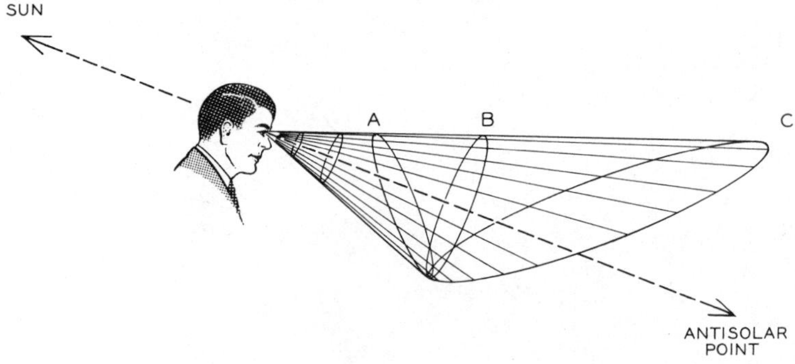

cloud. Like the glory, it varies in dimensions according to droplet size. It is bluish on the inside and reddish on the outside, unlike a halo, from ice crystals, which is reddish on the inside and bluish on the outside.

Since the center of a corona is the moon or sun and the center of a glory is the antisolar point, a glory is sometimes called an anticorona.

This chapter was to have been devoted to light. It inevitably had to deal with water droplets and ice crystals, clouds and mountains. The next chapter is labeled "Clouds," but some of it will deal with light. The world around us is not done up in mutually exclusive packages with labels like "Light" and "Clouds," "Physics" and "Chemistry," "Russian science" and "American science."

6. *Clouds*

A FEW MINUTES after takeoff in a fast jet plane you may hear a fellow passenger say, "I can't see anything any more — just clouds." Just clouds! "Did you ever think how unbelievable clouds would be if you had been born blind and no one had ever told you about them?" Guy Murchie asks the question in his sensitively perceptive book *Song of the Sky*. Or suppose you had the task of trying to make the blind person believe them. Try to describe them in all their shapes and sizes, colors and motions: the boiling black threatening thunderheads, the long level gray slivers against the sunset, the wind-torn high wisps made of ice crystals that Murchie calls "the silver eyelashes of the sun". From the ground we see them day after day, year after year, but still we pause now and then to admire them. From the air they take on new aspects, new colors, new shapes.

When we stood on the ground each white puffy cloud overhead seemed stepped down from the next nearest one until, far on the horizon, the clouds appeared to lie close to the ground. In Plate XXIV we see that this is an illusion. Flying up from the ground, we come abreast of such a cloud layer and, looking along it, see great numbers of clouds at the same

Plate XXIV. A layer of cumulus clouds viewed from (a) below, (b) alongside, (c) above

level (b, Plate XXIV). Rising above the cloud layer, we find that the perspective of our view is reversed. We now raise our eyes to see the far clouds (c, Plate XXIV). What causes these vast collections of puffy white clouds all lying in an orderly way at the same level?

On the ground we live at the bottom of the sea of air which coats the earth. The atmospheric load exerts a pressure of about fifteen pounds per square inch at sea level, but at higher elevations where the load of air above us is less, the pressure is less.

Air in contact with the sun-warmed earth is warmed, expands, and rises because it is less dense than the cooler air which has not expanded and which flows in from the side to take its place. As the rising air moves upward it is under less and less pressure and it therefore expands more and more. As a result of this expansion it gets cooler without losing any heat. This may sound impossible, but consider the diagram in Figure 18. The bottom box represents a certain volume of air at a given temperature. The heat that it has is a form of energy, the kinetic energy of all the air molecules dashing about in that volume of air and colliding with each other. The next box up, in Figure 18, represents the volume occupied by that same mass of air when it has risen into a region of less pressure. In expanding, this mass of air pushed back the air surrounding it, expending energy to do so, though not in the form of heat. As a

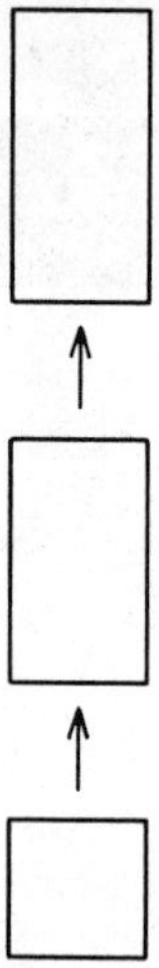

Figure 18. Diagram of a given mass of air expanding as it rises

result, the molecules now have less total kinetic energy, and the temperature is therefore lower. This process of becoming colder by expansion without losing heat is known as *adiabatic cooling*. The reverse process, becoming warmer because of compression, without gaining heat, is known as *adiabatic heating*. For each thousand feet such a parcel of air rises, the temperature falls about 5½ degrees Fahrenheit due to adiabatic cooling.

Warm air can contain more water in the gaseous state than cold air. At a lower temperature the water molecules dash about less vigorously and therefore have a greater tendency to stick to each other when they meet, that is, to condense into droplets of liquid water. A cubic meter of dry air at 30°C (86°F) can hold 30.4 grams of water as invisible water vapor, but the same amount of air at 6°C (43°F) could only hold 7.27 grams of water as water vapor. At 86°F, 7.27 grams of water would be only 24 percent of the amount the air could hold as water vapor. The amount of water vapor the air is holding at a particular temperature relative to the amount it could hold at that temperature is called the *relative humidity*. It is commonly expressed as a percentage. When the relative humidity reaches 100 percent the invisible water vapor in the air should condense into the visible water droplets which make up clouds. If the air is very clean this may not happen until even further cooling takes place because the droplets seem to need some particle to form on, a nucleus to initiate condensation. Laboratory experiments performed with filtered, de-ionized air, free of all particles except neutral molecules of oxygen, nitrogen, carbon dioxide and water, the constituents of clean air, have shown that in such specially prepared air relative humidity may reach several times 100 percent without condensation taking place.

In our atmosphere air is not that clean, especially at levels near the ground. From the evaporation of sea spray the air contains great numbers of salt particles, ranging in diameter from 0.01×10^{-4} to 10×10^{-4}cm., that is, from one millionth of a centimeter to one thousandth of a centimeter. In some places there are more than 1000 salt particles per cubic centimeter of air. These particles are especially important in initiating condensation since salt is hygroscopic, that is, water molecules adhere to its surface and it dissolves in the film of water and continues to attract more water molecules. This is why salt in a salt shaker gets sticky on humid summer days.

Many other kinds of particles can serve as nuclei for the condensation of water droplets. They are discussed in Louis J. Battan's paperback book, *Cloud Physics and Cloud Seeding* (New York, Doubleday, 1962.) Raindrops, forming on the particles in our atmosphere and carrying them to the ground, are air washers. If they did not do this, our air pollution problem would be even greater than it is.

In the upper atmosphere where nuclei are scarcer there are sometimes regions of high relative humidity in which no water droplets have formed. The air is crystal clear and full of water in the gaseous state. Into this saturated air roars our airplane. Whether it is a propeller plane or a jet, it spews out behind it the products of the combustion of its fuel, one of which is water vapor. Suddenly the air behind the plane has more water than before and plenty of nuclei for the water to condense onto as visible water droplets which scatter the light. In the wake of the plane a *condensation trail* forms which has acquired the nickname *contrail.* It may be formed of ice crystals, if the air is cold enough, as it often is, and it may stop abruptly when the plane flies into a region of lower humidity.

Sometimes the relative humidity may be so high that it takes no more than the decrease in pressure as the air passes over the wing to trigger the formation of water droplets. Then the wings wear condensation plumes.* The propellor tips on powerful prop planes sometimes generate air vortices with low-pressure centers which cause condensation if the relative humidity is high. When this happens, which is especially likely on take-off, a thin cylindrical shell extends backward from the propellor tips as the plane speeds forward.

When we discussed the shadow and glory phenomena in Chapter 5, we postponed mention of one until we had discussed its source. The contrail shadow can frequently be seen on the clouds below you. It lies like a long thin cigar with its forward tip always keeping pace with the plane and its back end spreading into a cloud farther back than you can see. At the front end there may be the bright spot mentioned in Chapter 5, and, surrounding the front end, the glory.

On the ground, we are used to seeing our shadows lengthen as the sun goes down, its rays becoming more nearly parallel with the surface of the earth where we stand. In the sky, as the sun's rays become more nearly parallel with the surface of the earth, shadows shift their positions to one more nearly on the same level as the object causing the shadow. In other words, as the sun sets, the shadow of the plane and its contrail rise. So, also, do the bright spot and the glory.

As you near any major airport, you are likely to see contrails of other planes that have just passed. They will be dense and

*For studying the formation of water droplets due to rapid decrease in pressure, C. T. R. Wilson in 1897 used a chamber in which he could cause a sudden expansion of the air. Such a chamber, now known as the Wilson cloud chamber, has been widely used for the detection of atomic-scale charged particles, since these nucleate condensation of water droplets under suitable conditions, and the track of the particle through the chamber is thus marked by a contrail.

compact, if recently formed, or open and larger if they have had time to spread out. Some of them may cast shadows which you can compare with the shadow of the contrail from your plane. Only by observing such a shadow can the crew tell when the plane has a contrail, a matter of no importance to them. They have no rear-view mirror.

The temperature of the air at which the invisible water vapor condenses to form visible liquid water is called the *dew point.* On a warm summer day, air in contact with an ice-cold glass is chilled, reaches the dew point, and some of the water vapor in it condenses into liquid water on the surface of the glass. In the atmosphere, adiabatic cooling of the rising air brings it to the dew point. Wherever the air is warmed near the ground, it rises and cools and where it reaches the dew point, clouds form.

When air moves over a mountain it has to rise and may reach the dew point with a resulting cloud forming just above the mountain (Plate XXV). Beyond the mountain the air moves down again, is warmed by adiabatic compression, and the water droplets in it evaporate. The cloud above the mountain hangs there, looking motionless, while, in the air

Plate XXV. Wave cloud on top of Mt. Shasta, California

that moves through it, water vapor condenses into droplets on the upwind side and within the cloud, only to evaporate into a gas again on the downwind side. Just as a submerged rock in a swift stream results in a hump of water over the rock and waves of decreasing amplitude downstream, so a mountain such as we have been describing produces not only the hump in the air flowing over the mountain, but waves of decreasing amplitude in the airflow downwind (Fig. 19). On the crests of each of these waves little lenticular clouds may form just as they did above the mountain, giving us visible evidence of the invisible waves in the sea of air (Fig. 19). The crest-to-crest distance, or wavelength, depends on the wind velocity and is generally between two and ten miles.

If the air flows over several mountain ridges, each may

Figure 19. Wave clouds

have a cloud above it and if the ridge-to-ridge distance happens to coincide with the airflow wavelength that would result from a single ridge, the maximum wave amplitude will result. Plate XXVI shows wave clouds over the ridges of the Appalachian Mountains in Pennsylvania. Because of the ideal conditions for their formation in this area, the Showers Project of the Department of Meteorology of Pennsylvania State University has made a study of them.

Cloud droplets are very small compared to raindrops. The maximum diameters of the droplets in cumulus clouds like those in Plate XXIV are less than 0.05 millimeters, commonly about 0.01 mm, and there may be as many as 300 droplets per cubic centimeter. Most raindrops are about a couple of millimeters in diameter.

As more water droplets accumulate the clouds grow bigger. Such clouds (Plate XXIV) are called *cumulus* clouds. The

Plate XXVI. Wave clouds over the Appalachian Mountains in Pennsylvania

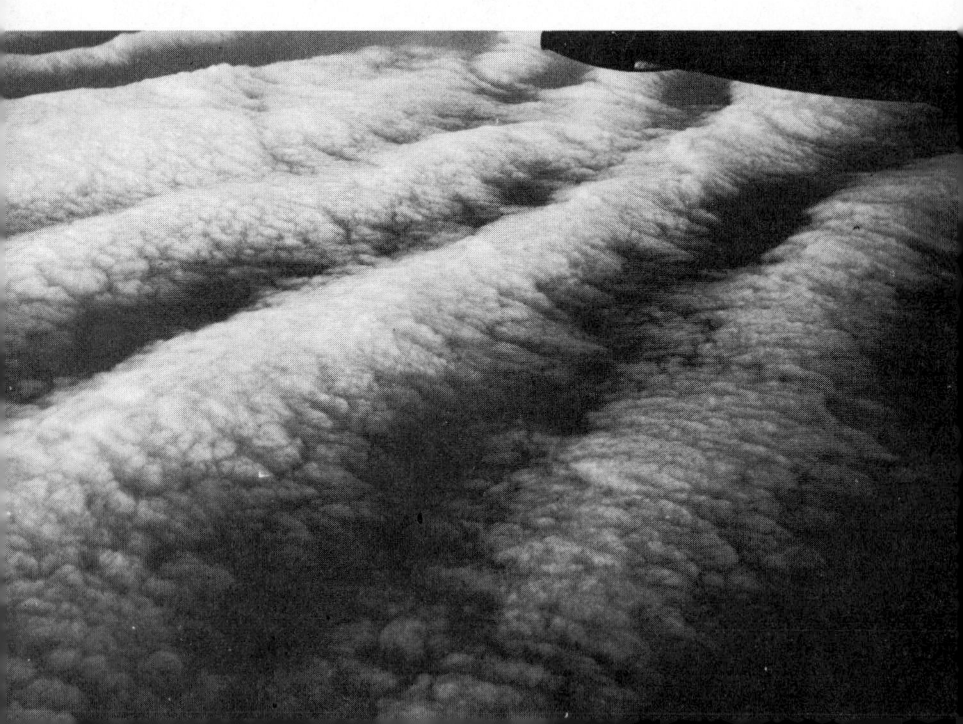

fact that they accumulate at about the same level over a wide area is an indication of the uniformity of temperature and relative humidity over that area.

Whether on the ground or in the air, you can watch this process taking place. Fix your attention on a small white cloud, low in the group. Remember as you watch it grow that there are about 300 droplets per cubic centimeter of growing cloud! A small cloud near the top of a cumulus layer may gradually dwindle and disappear.

The chart in Figure 20 will enable you to determine approximately the level of cumulus cloud formation for a particular set of ground conditions. For example, suppose the temperature on the ground is 70°F, and the relative humidity 50 percent. The point representing 70 degrees and 50 percent happens to lie on one of the curves in Figure 20, the curve for air containing 9.41 grams of water per cubic meter. Following this curve upward and to the right we find that this air would reach the dew point (100 percent relative humidity) if its temperature fell to approximately 50°F, a drop of 20 degrees. Since the temperature drops about $5\frac{1}{2}$ degrees for each thousand feet of rise, this air would reach the dew point if it rose to an elevation of 20/5.5 thousand feet or about 3600 feet.

Suppose the temperature on the ground is 75°F and the humidity 25 percent, represented by the point marked x on the 75-degree coordinate. Following the general trend of the other curves, we can see that the dew point of this air would be at about 36°F, or 39 degrees below surface temperature. If such air rises, is adiabatically cooled, and condensation occurs, the resulting cumulus clouds should form at 39/5.5 thousand feet or 7200 feet.

There are two modifying effects we have not taken into account in these calculations. In air at lower pressure the water molecules collide less frequently and so the dew point is lowered. The air would have to rise farther, because of this effect, before condensation occurred. On the other hand, we have neglected turbulence of the air which tends to promote condensation, thus modifying the effect of the lowered pressure. Due to the combination of these two effects our

Figure 20. Graph of relative humidity versus temperature for various amounts of water in air

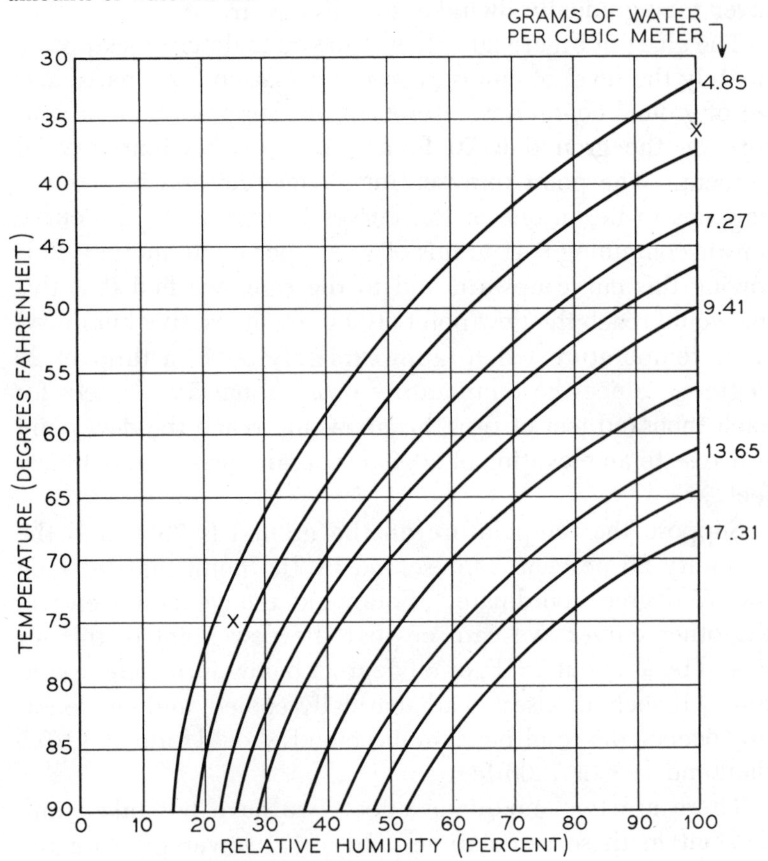

calculated elevations are likely to be a bit low, about nine tenths as great as the actual cloud elevation.

The fact that the very cold air of the upper atmosphere can hold very little water has turned out to be useful to airplane manufacturers. I discovered this as a result of a distressing incident on a recent trip. It was to be a long flight and before boarding the plane I carefully requested a seat assignment that would allow me to take photographs en route. I arrived on board with my camera only to find that the view outside my window was entirely obscured by water droplets which had condensed on the inside of the outer pane. One would suppose that the space between the two panes would be sealed off and that it would be filled with dried air before sealing so the cold of the upper atmosphere would not cause condensation. My seat companion volunteered the suggestion that the condensation would probably go away when we were in the air. It seemed obvious to me that this was nonsense, since the outer pane would be even colder when we rose to higher altitudes.

To my astonishment the prediction turned out to be true. I have since consulted an engineer with Douglas Aircraft on the subject of window construction, with very satisfying results. Both the outer window and the inner window in most planes are made of plastic approximately half an inch thick. The space between them is only an eighth to a quarter inch thick in some planes. The outer window is sealed at its edges, but the inner one is not, and air may gradually seep from the cabin into the inter-window space or from the inter-window space into the cabin. One would think that the breath of the passengers in the cabin air would cause condensation on the cold windowpane, just as it does in a car. But the air in the cabin is constantly being renewed with air from outside, air that has had most of the moisture wrung from it by the extreme cold. When it comes into the cabin and is warmed, its capacity to take up water increases greatly, that is, its relative humidity drops to a very low figure (follow one of the curves of Figure 20 from upper right to lower left). It is this thirsty air that seeps into the space between the windows when the plane is high in the air, drying up any water that may have condensed on the cold pane when the plane was on the ground.

Earlier in this chapter we saw how such condensation and evaporation in the undulating wind blowing over mountains can result in wave-clouds. Patterns of other sorts may occur under other conditions. Within a cumulus layer convection currents — the warm air rising and cool air descending — seem to form "cells," with the cumulus clouds capping the columns of rising air (Fig. 21). Experiments performed by many mete-

Figure 21. Air convection currents and cumulus clouds

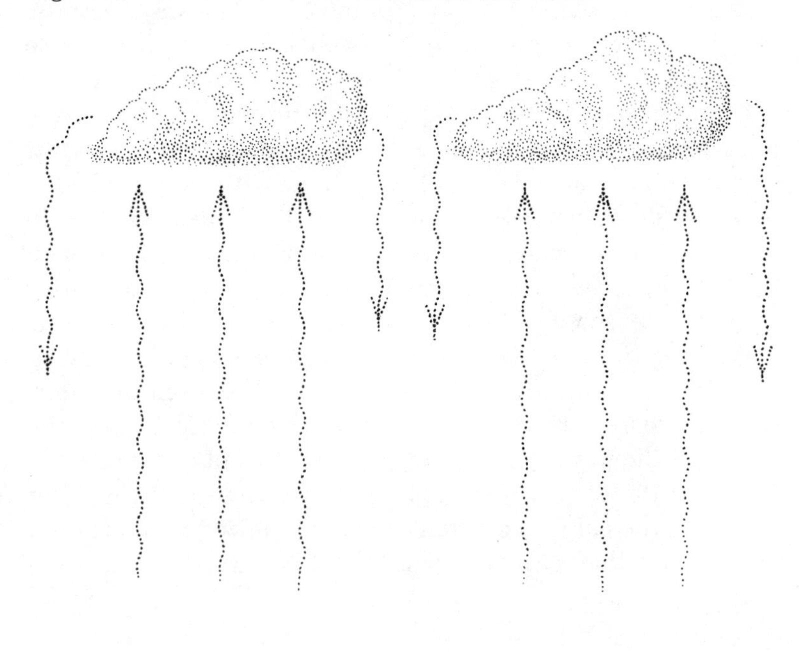

orologists in the laboratory have produced convection cells in a layer of liquid or in air between two horizontal plates, heated on the bottom. In one such experiment the plates were about a centimeter apart and the lower plate, made of brass, was heated to more than twelve Centigrade degrees hotter than the upper plate which was made of glass. Smoke was added to make air movements visible. The pattern of polygonal cells that forms in such an experiment, photographed through the glass plate, is shown in Plate XXVII. Each cell usually has five or six sides and is surrounded by other cells of the same sort. The diameter of a cell is about three times the distance between the plates. In this experiment the motion of the air is upward at the boundaries between cells, inward at the top and downward at the center of each cell where a dark spot may be seen in Plate XXVII. In this re-

Plate XXVII. Polygonal convection cells produced in a thin layer of smoke-filled air heated from below.

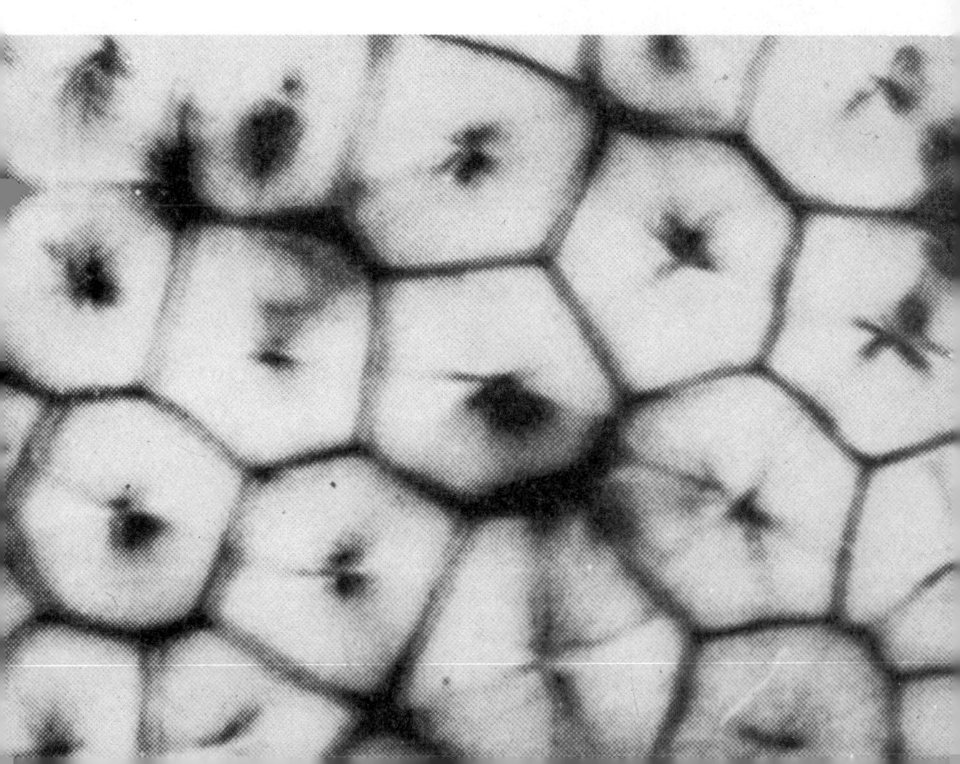

spect the behavior of these smoke clouds is unlike that of cumulus clouds in which the motion is upward in the center and downward at the edges. This experiment does serve to show, however, that under certain conditions of uniform heating from below, the convection currents in a gas will form a cellular structure.* The airplane passenger can observe a cellular structure of similar appearance in the cumulus layer below him (Plate XXVIII).

A plane flying beneath a layer of cumulus clouds, or through them, is alternately subjected to the updrafts and downdrafts of convection cells as shown in Figure 21. The flying is "bumpy" and the order is given to fasten your seat belts.

*For a description and photographs of these experiments, see *Tropical Meteorology* by Herbert Riehl (New York, McGraw-Hill, 1954), or the paperback *Clouds, Rain and Rainmaking* by B. J. Mason (Cambridge, England, Cambridge University Press, 1962).

Plate XXVIII. Cellular structure in a cumulus layer

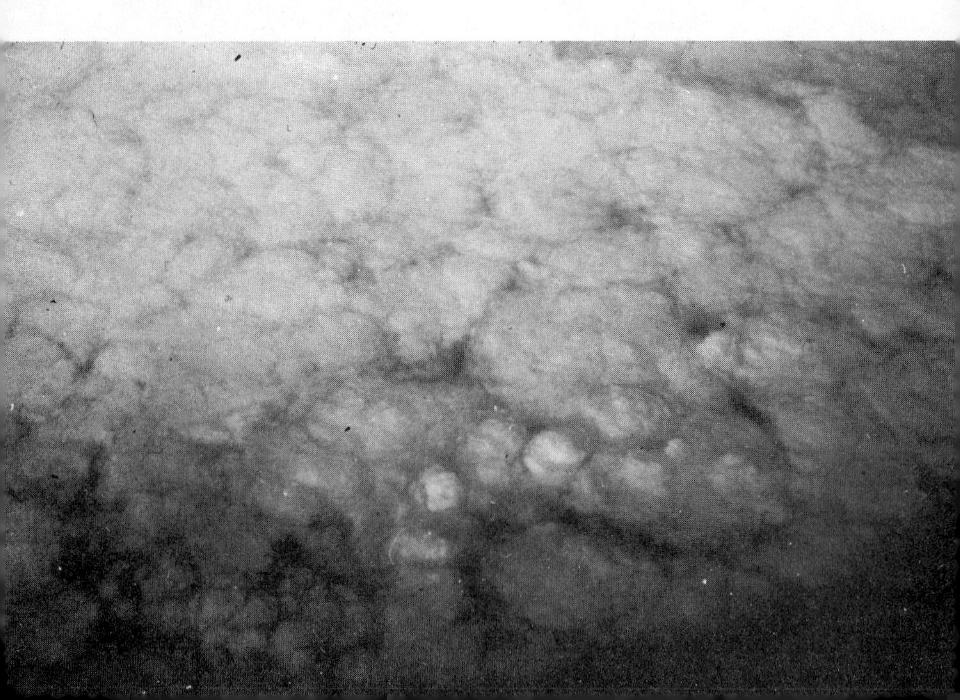

Above such a layer of clouds we no longer encounter convection currents, and the flying is smooth.

Why does the ascending air stop its ascent at the cloud? What determines the shape of the cloud in detail? We have seen that its level base occurs where the ascending air mass reaches the dew point, but what determines its bulging upper surface? There are several things going on at once in a cloud. As the water droplets in the cloud form from water vapor, heat is given off. This must be so, since heat was added to this water somewhere, originally, to change it into water vapor. (Recall the chilling effect of standing in a wet bathing suit from which water evaporates, getting the necessary heat for the process from your body.) Since heat energy, like any form of energy, can be neither created nor destroyed, it must be given off when the vapor changes its state to that of a liquid again. This released heat tends to increase the buoyancy of the air within the cloud.

The combined action of the rising air within the cloud and the descending air falling away along the sides, where evaporation of water droplets has taken heat from the neighboring air, gives the top of the cloud its rounded, cauliflower shape. In any one cloud there may be many local regions of rising and descending air as the upper surface engulfs pockets of the surrounding dry air. This gives the upper surface its many small bulges.

The warmth and moisture that the cloud contributes to its cooler dryer surroundings help to decrease the difference in density between it and the surrounding air. Since the buoyancy of the air responsible for the formation of the cumulus resulted from that difference in density, it will be lost when that difference ceases to exist. Old cumuli* lose their

* Meteorologists speak of a *cumulus* (noun) for a single cumulus cloud. The plural of *cumulus* is *cumuli.*

sharp outlines and often merge with each other in a shoddy agglomerate of stratocumulus form. If new warm currents of cloud-forming air rise into such an old-cumulus region, already full of water droplets, they will be protected by its dampness from arresting effects of evaporation that were just discussed. As a result they may form young cumuli which tower above the sea of old cumuli like those in the foreground of Plate XXIX. In the background of Plate XXIX are larger cumuliform clouds, most of which have the sharply outlined rounded tops typical of vigorously growing cumuli. The largest of these, however, has a flat top, which indicates that it has reached the level of its limit of rise and is no longer moving upward relative to its surroundings. The steady movement of the upper atmosphere has smeared it out into an anvil-like shape. Clouds of this sort can frequently be seen from a plane. Their level tops mark the *tropopause,* the upper limit of that region of the atmosphere in which the churning and mixing of warm and cool air, humid and dry air, results in our weather. This weather-producing region is called the *tropo-sphere,* from the Greek word *tropos,* meaning "turn." In Plate

Plate XXIX. The top of the troposphere at sunset

XXIX the tropopause is about at the level of the plane, about 34,000 feet. If the air is very moist cloud growth can continue to great heights, greater than 60,000 feet, and the updrafts may attain velocities of 4000 feet per minute.

At higher elevations the low temperatures result in ice-crystal formation though the temperature of the small water droplets may be cooled to far below 0°C before freezing takes place. The change from water to ice, like the change from gaseous water vapor to liquid water, is facilitated by the presence of particles. High above the earth there may not be enough dust to trigger the change of state, and water in small droplets may be *supercooled* water, liquid below its freezing temperature. The success of rainmaking by cloud seeding depends on the existence of supercooled water droplets which can be induced to freeze and fall.

Why should they fall when they freeze? If a droplet of water turned to a pellet of ice containing the same number of water molecules, it would weigh exactly the same but occupy more space in the sky because water expands when it freezes. It would require less updraft to keep it up than the water droplet would. However, at temperatures much below freezing, the tendency of water molecules to adhere to an ice pellet, making it large, is appreciably greater than their tendency to join a water droplet. Therefore when a tiny water droplet, too small to fall through the updrafts of air, is changed to ice, it quickly grows from a tiny ice pellet to one large enough to fall (one with a smaller ratio of wind-resisting surface to earth-attracted mass). When it falls into warmer air it melts again and falls as a raindrop.

Where the cumulus cloud droplets are changing to ice particles, the cloud boundary is less sharp. Some high ice-crystal clouds form directly from water vapor without the intermediate stage of liquid water. These are the *cirrus* clouds, high, white and wispy, which old weather-watchers used to call "mare's tails," clouds like the tail of a white horse, blown by the wind. These are usually at 25,000 feet or higher where the temperature is always far below freezing.

In the far background of Plate XXIX are some tenuous and diffuse cirrus clouds. These may be either the remnants of very old spread-out cumuliform tops or independently produced cirrus.

The layer-like *stratus* cloud is not formed as the cumuli are, but rather as fog is at ground level, merely by a decrease in the temperature of the air, in place, until the dew point is reached.

A young, newly formed cloud has droplets that are very small and fairly uniform in size. As time goes on these droplets come in contact with each other and coalesce to form larger droplets, some of which may get so large as to fall as rain. A big, dark rain-producing cumuliform cloud is called a *cumulonimbus* (examples can be seen in the middle distance of Plate XXIX) whereas stratiform clouds that produce precipitation are called *nimbostratus*.

Sometimes the influence of the temperature of the underlying land on the cloud patterns is sharply evident from the air. One day, on a trip from Albany, New York, to Newark, New Jersey, my plane flew just above a nearly continuous layer of cumulus clouds. Off to my right I noticed a wide gap in the clouds which continued for a long distance as though it had been plowed out by some giant snowplow. It was the region above the cold water of the Hudson River where the air was descending instead of rising and cumulus clouds were not forming.

The darkness or lightness of clouds is governed by their thickness, the size of the water droplets in them, the liquid-water content of the clouds and the light that falls on them.

A thick layer of stratus clouds will look dark to people on the ground. Very little of the light is absorbed by the clouds, but the amount reflected increases rapidly with their thickness. A layer of clouds 300 feet thick will reflect back about 35 percent of the sun's light falling on it. From the plane we look down on the bright white reflecting surface of the clouds and receive the light that is not reaching the people on the ground. The reflectivity is greater for clouds with high water content and is greater when the cloud droplets are small than when they are large. Some clouds are dark because they are in the shadow of other clouds (Plate XXIX).

A scene such as that in Plate XXIX tells a rich and complex story to the meteorologist capable of interpreting it in detail. This photograph was presented to two such meterologists. The first said it would take a whole day to analyze the information in the picture. The second said it would take a month. The analysis would fill a larger book than this.

Sometimes, in some places, the churning and mixing that give the troposphere its name cease, and tranquillity prevails in the lower atmosphere. For this to occur, the air on the bottom must be heavier than the air that overlies it so that it has no tendency to rise into the upper atmosphere. This will happen whenever conditions are such that the air near the surface of the earth is cooler than the overlying air. Since this is the reverse of the usual condition, it is called a *temperature inversion.* Color Plate II, on the inside front cover, shows what it looks like from the plane. The hazy, heavy motionless air of the lower atmosphere lies like a thick blanket near the ground, while above it the clear, lighter sun-warmed air rests lightly on its sharply-defined surface. The lower air is filled with more than haze. From the thousands of cars of the greater New York area, from power plants and factories and chemical plants of all sorts, waste products are poured out into the air. When the wind is blowing or when the usual convection currents are operating and the warm air is rising into the cooler upper atmosphere, this atmospheric refuse is carried away. When a temperature inversion takes place, the city sits in a pool of its own waste, risking death by asphyxiation.

7. *Wind, Rain and Snow*

"We are cruising at an elevation of 28,000 feet with a ground speed of 420 miles per hour. We have a head wind of 150 miles per hour. Our arrival at Minneapolis will be on schedule."

The passengers listen and return to their reading. Nobody is surprised. Yet in 1933, A. K. Lobeck wrote, in *Airways of America*:

> The prevailing westerly winds blowing across the country have an average velocity near the surface of the earth of between 5 and 15 miles an hour in different parts of the country, being greatest in the Great Plains and least in the basin areas between the mountain ranges of Wyoming and of central California. . . . At higher altitudes the velocity is much greater, hence the practice of flying eastward at altitudes of 10,000 feet and westward at only 1,000 to 2,000 feet above the surface of the earth.
>
> The contrast between eastward and westward flying may be best appreciated by considering the effect of only a 10-mile wind, during a flight of 5 hours. On the eastbound trip this means a tailwind which adds 50 miles to the distance travelled in 5 hours. On the westbound trip this is a head-

wind which reduces the distance travelled by that amount. The total difference between the two flights is 100 miles, which is the best part of the distance flown in one hour. A wind of 20 miles per hour is not at all uncommon. This means almost two hours difference in flying time. Or, if we consider a run of 400 miles with a wind of 40 miles an hour, we find that it takes almost twice as long to travel this distance against the wind as with the wind. It is not difficult to understand, therefore, why air schedules cannot be adhered to so rigidly as railroad schedules.

Much of this is just as true today as it was in 1933. A 10-mile-per-hour head wind adds 10 miles of flying distance during every hour of flight. The plane is in a treadmill of air. Its motion ahead is relative to the air in which it moves. With a 150 mph head wind on a two-hour stretch of flight, the plane travels through 300 more miles of air during those two hours than it would have traversed had the air been still.

How, then, do airplanes keep to their schedules today when they couldn't thirty years ago? First, airlines have benefited from thirty years of experience with the winds. The timetables allow about five hours for the flight with the wind from San Francisco to New York, but about six hours for the flight against the wind from New York to San Francisco. Second, today's planes fly faster. A 60 mph head wind reduces by 60 percent the speed of a plane going 100 mph, but reduces by only 10 percent the speed of a plane going 600 mph. The plane bound for Minneapolis probably did not have the 150 mph head wind for long. Such wind velocities are rare except in *jet streams,* those rather well-defined, flat-tubular rivers of air which follow a wandering, shifting course across the continent. The maximum wind velocity is in the core of the stream and may exceed 250 mph. A pilot who finds himself flying upstream in a jet stream will arrange to get out of it as soon as possible.

The speed of today's jet planes, relative to the air, may be close to that of sound in air. At these speeds a *shock wave,* a jump in air pressure and density, spreads outward and backward from the nose of the plane. Since such a pressure disturbance would produce an undesirable drag on the plane as it passed by the wings, it is advantageous to have the wings "swept back" at such an angle that they keep out of the wake-like path of the disturbance. However, this is not the only factor that must be taken into account in their design.*

Airplanes traveling at such speeds are traveling in the stratosphere, that relatively nonturbulent layer of atmosphere above the churning, weather-making troposphere which is next to the earth's surface. In the stratosphere there is no rain or snow. Only occasionally does an energetic thundercloud rear its head up through the tropopause that marks the limit of the troposphere. Just below this limit we commonly see the vast, continuous layer of cloud tops below which weather is taking place.

As we go down into the troposphere we feel the turning and turbulence of air that gave it its name. If the clouds are thick the day will be dark beneath them and precipitation may be taking place. The tracks of raindrops on the windows of the plane are not vertical tracks as they are on the windows of a house. Their orientation depends on the vector sum of the motion of the raindrops and the motion of the plane. Which is the larger of these two vectors?

The combined effects of the vibration of the airplane, the various directions and velocities of air sweeping past the outer surface of the window, the gravitational attraction between the earth and the raindrops, and surface tension (acting always like a tight surface skin on the water droplets) result in a great

*See P. L. Sutcliffe's discussion of wing angle versus speed in *Supersonic Engineering,* edited by J. T. Henshaw and published by Heinemann, London, 1962.

variety of phenomena. The surface of the window does not "wet" very well; that is, water does not spread out on it in a uniform film. If a streak of water forms as a raindrop strikes the window, it quickly breaks up into many drops because of surface tension. As the drops grow with added rain, two may touch and quickly flow together to make one big drop with less surface area than that of the two separated drops. If you are standing on the ground this drop may start to move down the window, incorporating other drops as it goes, and leaving behind it a streak of water which quickly breaks up into small drops again.

When a propeller-driven plane is standing on the ground with its motors idling, the raindrops on the window behave as they would on the window of a house, but when the propellers are rotating at high speed with the plane still on the ground a very curious phenomenon is observable. As the air moved by the propellers blows back along the windowpane, the drops move in jumps, not smoothly across the pane. All the drops jump at once, hesitate a moment and then jump again, all together. What is the explanation of this simultaneously jerky motion of the drops? Is the air from the propellers so sharply gusty at the windowpane surface that it blows and stops, blows and stops within such a short period? Does the vibration of the windowpane periodically result in reduction of the force holding the drops against the pane? Further research is needed.

When the sun shines onto snowflakes in the air below you, it sometimes produces a well-defined bright round spot which seems to race over the ground at the speed of the plane. The reflection of such a sharp image of the sun by air-borne snow shows that the flakes must be flat platelets, most of which are floating horizontally so that they act as millions of tiny horizontal mirrors.

If a light snow has fallen you may be able to see something that can be observed best (perhaps only) from the air under these conditions. In vacant lots, in fields and sparse woodlands, wild animals make countless little tracks, crisscrossing in random ways that suit their small purposes. When a light snow has fallen, these tracks show as a white network in the brown vegetation.

Snow may last in some places and melt away in others because of differential heating. The cause of the difference may be obvious, such as exposure to sun or protection from it, or it may be more subtle. I know of one case in which the snow over a newly laid pipe remained when the snow on either side had melted away. Perhaps the trapped air in the loosened soil served to make it a better insulator, a poorer conductor of the earth's warmth which melted the snow on either side.

The distribution of snow in the early stages of a snow storm when the ground is not yet all white may be very different from the distribution a few days after a snow storm when some of the snow has melted and again the ground is not all white. In the early stages of a light snow, the boundaries of fields may show less snow because the dark rocks of the stone walls absorb more of the sun's heat than the fields do and so cause melting of the snow. A few days after a heavy snow the fields may be edged in white where the snow drifted deep against the stone walls and is therefore taking longer to disappear.

Small lakes may remain frozen when the ice on large lakes melts. Indeed small lakes generally freeze before large lakes because a large body of water has a greater heat capacity than a small body of water does and takes longer to lose this heat. Lake water loses heat largely from the surface, especially on a clear night, when heat is radiated to the cold sky. The cold surface water is denser than the warmer underlying water and therefore sinks to the bottom of the lake leaving warmer water to be cooled at the surface.

When ice forms it does not sink to the bottom. As we have seen, it stays on the surface because it is less dense than water. The ice layer itself is an insulator, cutting down the loss of heat from the surface of the lake and the freezing of a lake is thus a self-limiting process.

A thicker snow provides some contrasts that sharpen our observation of the land below us. A smooth white blanket marks the cultivated land between the brown areas of trees and shrubs. One time, as I flew over the geologically young region near Pittsburgh, where the snow-white cultivated areas on the flat divides were separated by the narrow tree-brown valleys, it seemed to me that the land resembled a pan of sugar-powdered buns.

When there is snow on the ground, the wooded white hills are clearly outlined in brown against the hills behind, since we look through many trees as we look along the side of a hill, but through few as we look at the face of a hill. This is the crew-haircut phenomenon. Unless a man's hair is very thick, you can easily see the skin on his head when you look down on his crew haircut from the top. Looking at him from the side, however, you see a fine stand of bushy hair.

In barren country the snow-laden ledges contrast with the dark steep slopes so that a hard-rock layer which might otherwise escape notice stands out sharply marked in white. Lakes and streams, if they are not covered with ice, become the dark features of the snow-covered landscape.

A part of the land that is familiar to you from the air becomes a different thing after the snow, with some of the old features hidden, but with new ones brought to light.

What of the work of the wind on the land? Can any of

it be seen from the air? The shapes of sand dunes may be distinctive from above. In semi-arid regions, where farming is possible, with some irrigation, a part of the land that is higher and therefore drier may develop dunes if the soil is sandy and the vegetation sparse. Just east of Denver patches of parabola-shaped dunes may be seen, close to orderly cultivated fields. These parabolic, or bowout, dunes are typical of semi-arid plains and plateaus where vegetation exerts some influence.

In areas where the sand movement is not hampered by vegetation, a different type of dune may develop. Where the wind is usually from a particular direction, the dunes may "march" across the land, developing crescent shapes as they go, with horns pointing away from the wind. The center of the due is an obstruction over which the wind must rise, dropping sand all along the gentle upwind slope of the dune. Some of the wind sweeps past the ends of the dune, carrying sand farther along to form the horns of the crescent. Such dunes are called *barchans* and may be seen from the air when you cross the Mojave desert.

8. *Quantitative Measurements*

PLANES fly in lanes. Cars on the ground are restricted to roads and must keep to the right. For these we need only two-dimensional regulations because the force of gravity regulates their motion in the third dimension. For planes there are three-dimensional regulations. They must fly at flight levels specified by the Federal Aviation Agency. In general, planes flying west fly at even thousands while planes flying east fly at odd thousands of feet elevation above mean sea level, but special conditions may require departure from this rule.

Commonly the captain announces to the passengers the elevation at which the plane is flying. This information, together with your bookback goniometer, makes it possible for you to determine the dimensions of lakes, rivers, fields and large features of the landscape below you, as well as how far away they are.

Referring to Figure 22, we see that if you hold the long edge of the book (G) as nearly vertical as possible* and sight

*The top of the book will thus be horizontal at eye level. Sight along the top edge to check this.

along the line labeled *e* to some distant object, then the distance (*e*) from your eye to that object can be found from the known elevation of the plane (*b*) by measuring the angle A, since

$$\cos A = \frac{b}{e}$$

$$e = \frac{b}{\cos A}$$

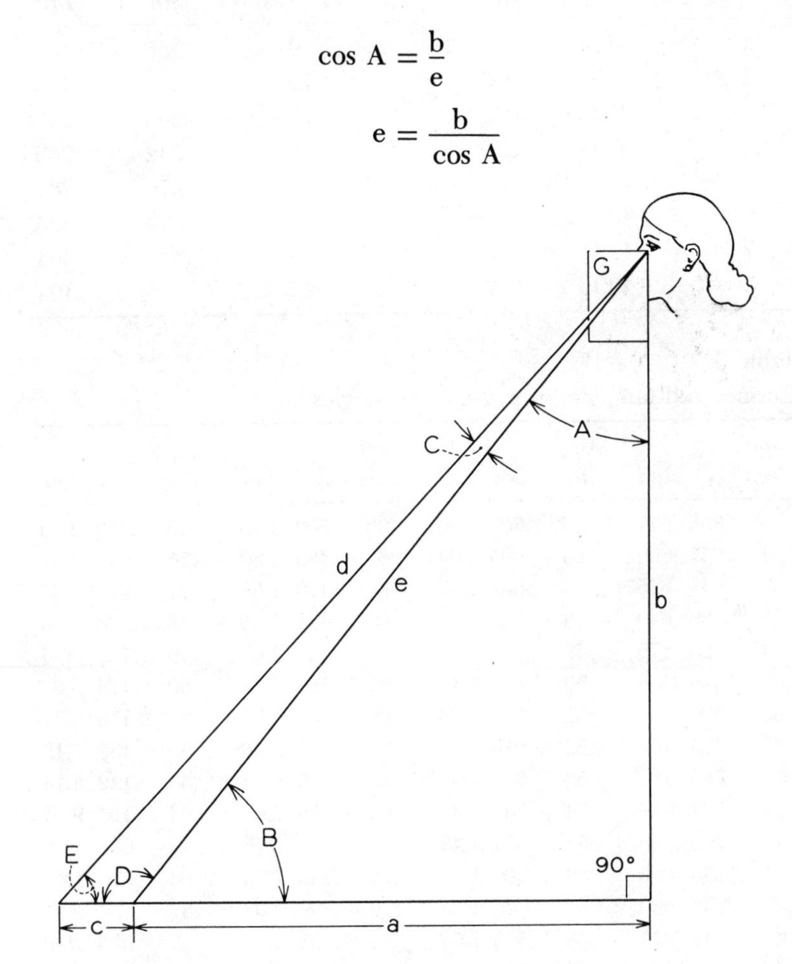

Figure 22. Use of the bookback goniometer to make ground measurements

Table 2
Sines and tangents of some small angles

Angle (degrees)	*Sin*	*Tan*	*Angle (degrees)*	*Sin*	*Tan*
1	.017	.017	9	.156	.158
2	.035	.035	10	.174	.176
3	.052	.052	12	.208	.213
4	.070	.070	14	.242	.249
5	.087	.087	16	.276	.287
6	.105	.105	18	.309	.325
7	.122	.123	20	.342	.364
8	.139	.141	22	.375	.404

Table 3
Cosines and tangents of some larger angles

Angle (deg.)	*cos*	*tan*	*Angle (deg.)*	*cos*	*tan*	*Angle (deg.)*	*cos*	*tan*	*Angle (deg.)*	*cos*	*tan*
30	.866	.577	45	.707	1.00	60	.500	1.73	75	.259	3.73
31	.857	.600	46	.695	1.04	61	.485	1.80	76	.242	4.01
32	.848	.625	47	.682	1.07	62	.470	1.88	77	.225	4.33
33	.839	.649	48	.669	1.11	63	.454	1.96	78	.208	4.70
34	.829	.675	49	.656	1.15	64	.438	2.05	79	.191	5.14
35	.819	.700	50	.643	1.19	65	.423	2.14	80	.174	5.67
36	.809	.727	51	.629	1.23	66	.407	2.25	81	.156	6.31
37	.799	.754	52	.616	1.28	67	.391	2.36	82	.139	7.12
38	.788	.781	53	.601	1.33	68	.375	2.48	83	.122	8.14
39	.777	.810	54	.588	1.38	69	.358	2.61	84	.105	9.51
40	.766	.839	55	.574	1.43	70	.342	2.75	85	.087	11.4
41	.755	.869	56	.560	1.48	71	.326	2.90	86	.070	14.3
42	.743	.900	57	.545	1.54	72	.309	3.08	87	.052	19.1
43	.731	.933	58	.530	1.60	73	.292	3.27	88	.035	28.6
44	.719	.966	59	.515	1.66	74	.276	3.49	89	.017	57.3

Tables of cosines of angles as well as sines and tangents are given in this chapter for your convenience.

You can also find the ground distance (a) from a point directly beneath the plane to the object sighted, since

$$\tan A = \frac{a}{b}$$

$$a = b \tan A$$

Suppose the distance c (Fig. 22) is the width of a lake to be measured in the same direction as that in which the distance a was measured. You can determine this by measuring angle C (as well as A), and by knowing e.

From "the sine law" which holds good for all triangles, regardless of shape,

$$\frac{e}{\sin E} = \frac{c}{\sin C}$$

$$E = 90 - (A + C)$$

so

$$\sin E = \cos (A + C)$$

and

$$c = \frac{e \sin C}{\cos (A + C)}$$

To measure the length of the lake that is parallel to the route of the plane, determine the distance from your eye to the lake (e) first, and then turn the goniometer so that you sight along the spine of the book to one end of the lake and along some line of the goniometer to the other end of the lake. Record the angle (S) between the two lines of sight and find the length of the lake (s) in this direction from

$$\tan S = \frac{s}{e}$$

$$s = e \tan S$$

If you do not know your elevation, you can determine it by using the bookback goniometer if you know the dimensions of some feature on the ground. Roads and houses and trucks and cars, all of whose dimensions we know and might think we could use, are too small for this purpose. One needs to carry a map. Automobile road maps for each state are about the right scale. Shapes of lakes and rivers, parks and road patterns on the map can be recognized on the ground and used for measurement since the map carries a scale.

Referring again to Fig. 22, if the dimension (c) of some feature on the ground is known (measured in the direction of the distance a) you can determine your elevation (b) by measuring angles A and C since, as we found before,

$$c = \frac{e \sin C}{\cos (A + C)}$$

and therefore

$$e = \frac{c \cos (A + C)}{\sin C}$$

But

$$e = \frac{b}{\cos A}$$

So

$$b = e \cos A = \frac{c \cos (A + C) \cos A}{\sin C}$$

If you measure the angle S, subtended by the length of the lake (s) parallel to the route of the plane, as we did before, then, by knowing s from the map, you can find e from

$$\tan S = \frac{s}{e}$$

$$e = \frac{s}{\tan S}$$

and then we have the very simple equation

$$b = \frac{s \cos A}{\tan S}$$

Since landscape features move past the window rather quickly, you may wish to use a faster method for making measurements. If you carry a writing pad with you, you can use it just as you would the goniometer, making marks on the far edge of the pad to indicate your line of sight (Fig. 23a), and measure the angles later by folding the paper along the line of sight and matching the angle on the bookback goniometer (Fig. 23b).

Figure 23. Use of a writing pad for faster recording of angles

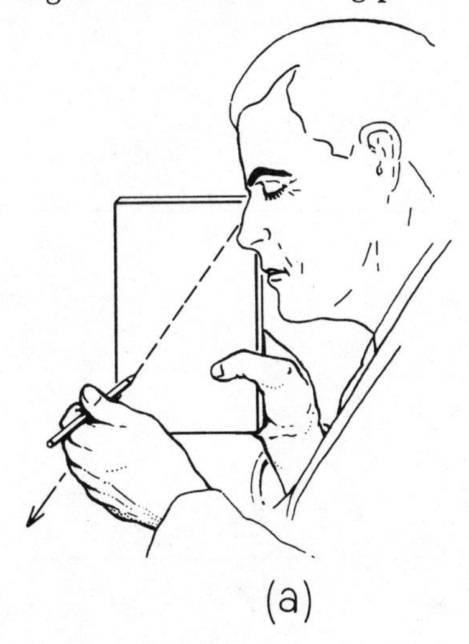

(a)

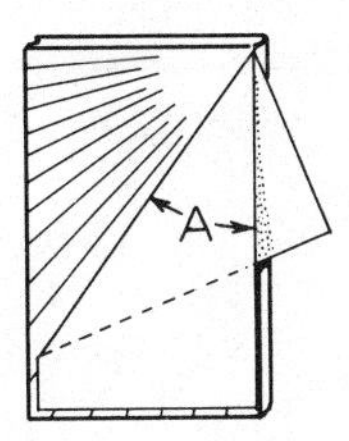

(b)

For a quick, rough estimate of your elevation when you are near the ground, you might see how much of the shadow of the airplane is covered by your little finger. Since the shadow is made by the sun's rays, which are approximately parallel, it will be the same size as the airplane while the plane is near the ground. If you know the size of the air-

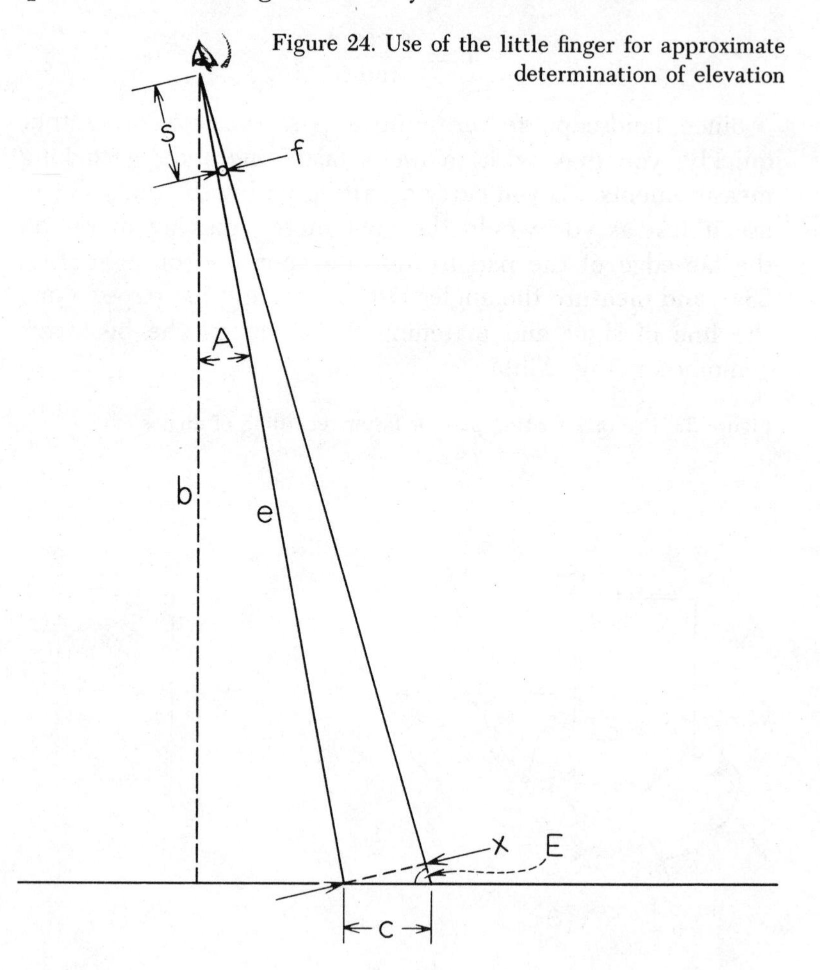

Figure 24. Use of the little finger for approximate determination of elevation

plane, the diameter of your little finger and the distance of your little finger from your eye, you can make a rough determination of your elevation (Fig. 24).

By similar triangles,

$$\frac{e}{x} = \frac{s}{f}$$

$$e = \frac{xs}{f}$$

For example, if the plane is 100 feet long and its shadow is just covered by your half-inch-thick little finger, held 8 inches from your eye, then $x = 100$ ft., $s = 8$ in. $= \frac{2}{3}$ ft., and $f = \frac{1}{2}$ in. $= \frac{1}{24}$ ft.

$$e = \frac{100 \times \frac{2}{3}}{\frac{1}{24}} = \frac{4800}{3} = 1600 \text{ ft.}$$

If the sun is so high that the angle E, Figure 24, is large, the distance x can be taken as approximately equal to c, the size of the plane's shadow and therefore of the plane. Similarly e will be approximately equal to b, your elevation. More precisely, $x = c \sin E$ and $b = e \cos A$.

To use this rough method you must know the distance between your eye and your little finger. There is seldom room to hold it at arm's length, a distance you could easily determine. A convenient means of measurement would be to place your thumb against your nose and stretch your little finger out as far as possible. The span of your hand is probably about 8 inches and could easily be determined. However, this arrangement might attract unfavorable notice from your fellow passengers. It would be better to use the book as a spacer between your eye and your little finger. It is just $8\frac{1}{4}$ inches long including the cover.

As we watch the shadow of the plane race across the

ground, we can make a rough comparison between its ground speed and that of cars or trains. On the sunny side of the plane we can use the rate at which the point of maximum reflection from water surfaces moves across the land if the plane's route is roughly at right angles to the sun's rays. The sun is so far away that all its rays reaching some part of the earth may be considered approximately parallel to each other.

Figure 25. Use of flash reflections from water to determine the speed of the airplane

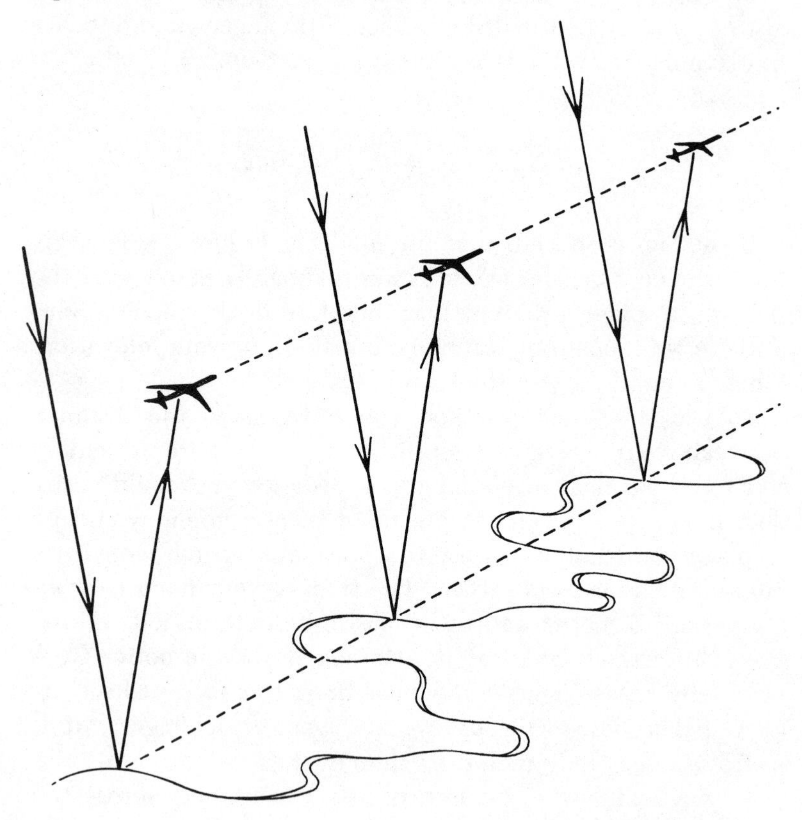

In Figure 25 we see that the point of land from which a horizontal reflecting surface will reflect the sun's rays into our eyes in the plane will move with the same velocity as the plane. If we know from a map the distance between two reflecting points on land and can measure the time between the flash reflection from one and the flash reflection from the other, we can determine our cruising speed.

What is the speed of the plane as it lands? Color Plate III, on the inside back cover, is a time exposure of an airplane landing at Newark Airport at night. The picture was taken from the control tower. In the background is the tower of the Empire State Building. The bright curve is due to the landing lights of the plane, the broken red streak above this curve is from the flashing red light which tops every airplane. On the ground, yellow-white lights mark the sides of the landing runway. If we knew how far apart the runway lights are and how frequently the red light on the plane was flashing, we could determine the speed of the plane as it was landing. (Near the right side of the picture, the red spots from the flashing light appear closer together because the path of the plane was more directly toward us as it circled for a landing. Along the runway its path must have been parallel to the runway since no disaster occurred.)

These particular runway lights are 200 feet apart. Federal Aviation Agency regulations require that the flashing red light on the top of the plane shall flash no less than 40 times a minute and no more than 100 times a minute. So we do not know the frequency of the red flashing light. Let us suppose that on this particular plane it was flashing 60 times a minute, a figure well within the regulations and one which makes our calculations easy since the distance from one break in the red streak to the next break would then represent the distance traveled in one second. How far did the plane travel in three seconds? How fast was it going in miles per hour?

In this problem we have had to *assume* a particular flashing frequency for the red light, in order to have a time measurement. You may be able to measure the speed of your plane as it lands if you have a watch with a second hand and can count how many runway lights go by your window in, say, five seconds.

The standard distance between lights along the edge of a landing runway is 200 feet, as in Color Plate III where the plane is landing on the east-west runway at Newark. However, the northeast-southwest runway at Newark and the east-west runway at Philadelphia are marked by lights 100 feet apart.

Sometimes the captain will announce the ground speed of the airplane and then you can determine distances between objects on the ground by timing the interval between flash reflections from adjoining water bodies or, on the other side of the plane, superposition of the plane's shadow over the objects. A captain may announce the speed in knots since airlines actually use this unit. A knot is not a unit of distance, it is a unit of speed. One knot equals one nautical mile per hour and a nautical mile is the length of one minute ($1/60$ of a degree) of longitude at the equator, which makes it equal to 1.15 statute miles (the mile that is 5,280 feet long). A speed of 100 knots is a speed of 115 miles per hour.

It is a little surprising that we have not invented a single speed unit to replace the wordy "miles per hour" that we use so often. We have a single name for a unit of work done in a given length of time: the watt. When the electric company bills you for the work done by their electricity they have to multiply the work-per-unit-time unit by the time.

$$\frac{\text{work}}{\text{time}} \times \text{time} = \text{work}$$

$$\text{kilowatts} \times \text{hours} = \text{kilowatt hours}$$

They cannot bill you for the electricity, since all of that comes back to them through the circuit; they properly bill you for the work it did while it was on its way.

What would you multiply a knot by to get a nautical mile?

For the measurement of distances in Figure 22, we assumed that the surface of the earth was flat. When our elevation is not very great, this approximation serves reasonably well, but when we fly high, the horizon gets farther away because a straight line of sight extends farther over the bulge of the earth (Fig. 26). We can easily calculate the relationship between the distance to the horizon and our elevation if we make some simplifying assumptions.

Figure 26. The higher the airplane the farther the horizon

In Figure 27 we have assumed the earth to be as smooth as a billiard ball with its surface at mean sea level. The (greatly exaggerated) elevation of the plane above this surface is *b*, the distance to the horizon is *e* and the radius of the earth (3,960 miles) is *r*.

From the Pythagorean Proposition,

$$\begin{aligned} e^2 + r^2 &= (r+b)^2 \\ e^2 &= r^2 + 2rb + b^2 - r^2 \\ &= 2rb + b^2 \end{aligned}$$

The quantity b is so small compared to r that the value of b^2 compared to $2rb$ is relatively so extremely small that it can be neglected.

$$\therefore e \cong \sqrt{2rb}$$

$$\sqrt{2r} = 89.0$$

So

$$e \cong 89\sqrt{b}$$

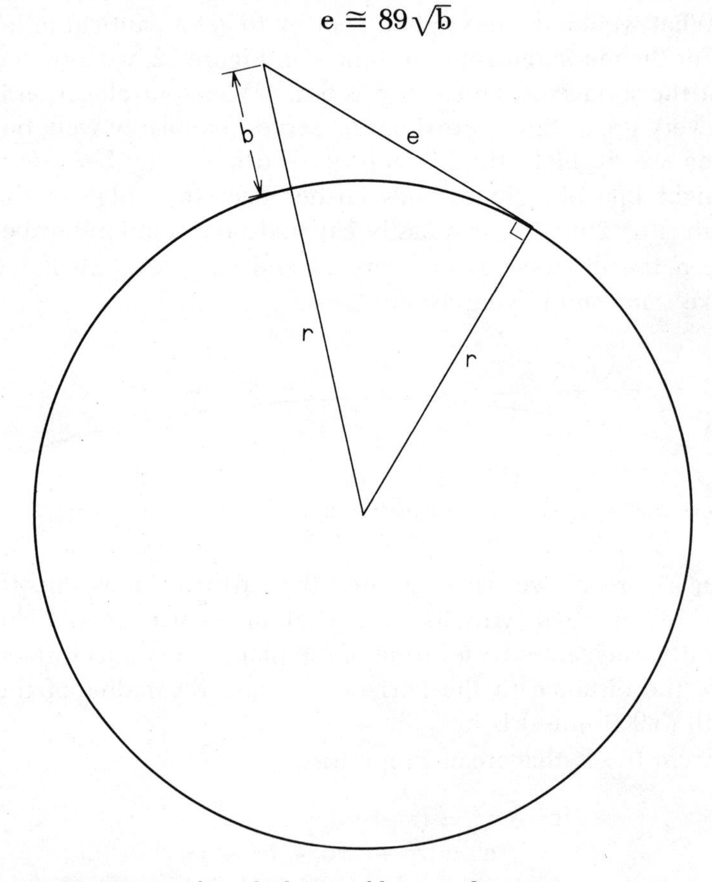

Figure 27. Diagram for calculation of horizon distance

For example, suppose your elevation, b, is 31,000 feet or 5.87 mi.

$$\sqrt{5.87} = 2.42$$

The distance to the horizon is

$e \cong 89 \times 2.42 = 215.7$ mi. or approximately 200 miles.

Alternatively we could write

$$\begin{aligned} \text{e (in miles)} &= \sqrt{\text{b(in feet)}} \times \sqrt{1/5280} \times 89 \\ &= \sqrt{\text{b(in feet)}} \times 1.23 \quad \text{(easy to remember)} \end{aligned}$$

Example:

$$\begin{aligned} e &= \sqrt{31{,}000} \times 1.23 \\ &= \sqrt{310} \times \sqrt{100} \times 1.23 \end{aligned}$$

Now a convenient trick for getting the square root of a number, n, such as 310, is to take out of it a perfect square, x, such as 289 (17×17). Then

$$\sqrt{n} \cong \sqrt{x}\left(1 + \frac{n - x}{2x}\right)$$

$$\begin{aligned} \sqrt{310} &\cong 17\left(1 + \frac{310 - 289}{2 \times 289}\right) \\ &= 17\left(1 + \frac{21}{578}\right) \\ &\cong 17 \times 1.04 = 17.68 \\ e &\cong 17.68 \times 10 \times 1.23 = 217 \text{ miles} \end{aligned}$$

This is the same as the previous answer within the limits of error introduced by rounding off numbers.

The relation $e \cong 89\sqrt{b}$ is plotted in Figure 28, with the elevation (b), in feet, plotted along the ordinate and the distance to the horizon (e), in miles, plotted along the abscissa.

The relationship does not give true horizon distances very exactly for two reasons. First, the surface of the earth has humps and hollows which we have disregarded. The peaks of distant mountains will be visible beyond the horizon we have calculated. Second, the path of a light beam coming to us from distant objects is bent as it goes through air of different densities. The beam is bent downward toward the earth as it travels up through the atmosphere passing from denser to rarer air. This makes it possible for us to see over the curve of the earth a bit farther than we could see if our line of sight were perfectly straight as we assumed it to be in our calculations.

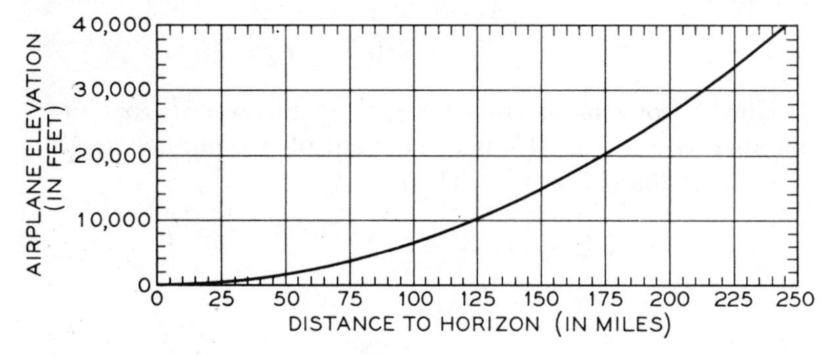

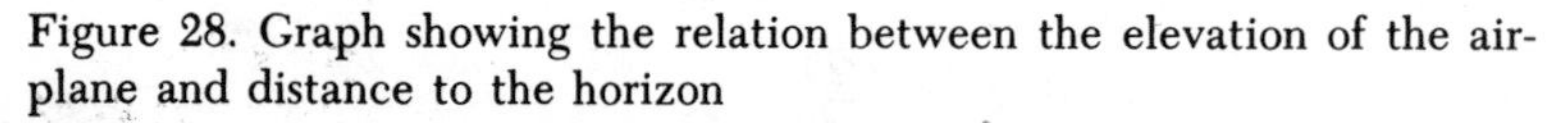

Figure 28. Graph showing the relation between the elevation of the airplane and distance to the horizon

An interesting result of the relation $e = 89\sqrt{b}$ is that the *area* visible from the plane is (approximately) simply proportional to the altitude.

The area of our circular field of view is $A = \pi r^2$ where r in this case is approximately the distance e which is equal to $89\sqrt{b}$. So

$$A \cong \pi \times (89)^2 b$$
$$A \cong 25{,}200 \text{ b}$$

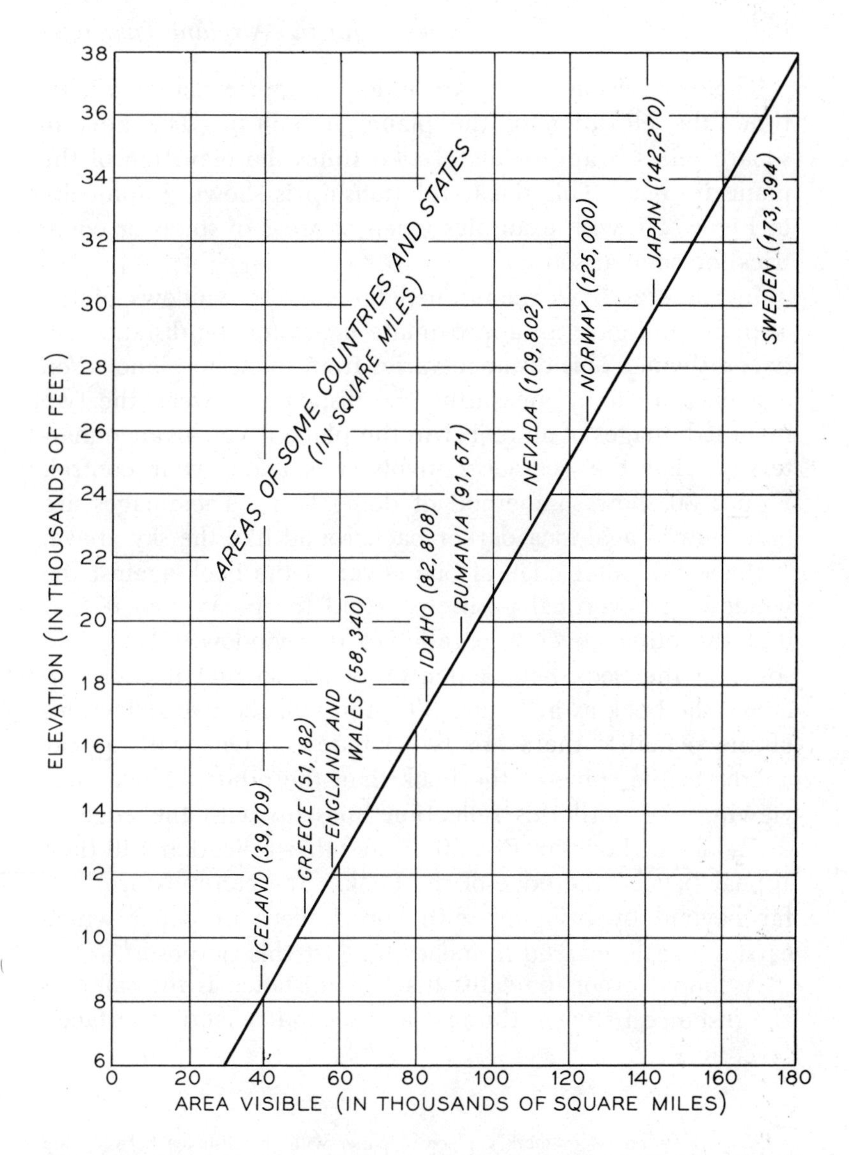

Figure 29. Graph showing the linear relationship between the elevation of the airplane and the area of the land visible

The area visible (in square miles) is approximately 25,200 times the elevation of the plane (in miles). The area in square miles is approximately 4.8 times the elevation of the plane in feet. This linear relationship is shown graphically in Figure 29, with examples given of areas of states or countries for comparison.

In Chapter 6 we mentioned the double windows of the airplane and gave an approximate figure for the distance between them. This distance varies from plane to plane. You can measure it by measuring the distance between the two reflected images of an object in the plane if you arrange matters so that the geometry involved is under your control. Figure 30 shows a method of doing this. The images are best viewed against a darker background like the sky above.

Place the outer edge of one cover of the book against the window in a vertical position. Hold the book open 90° so that the other cover is parallel to the window. Along the spine of the book hold a pen or pencil so that it sticks up above the book as in Figure 30. You will see two reflections of the pencil if there are two windows. One will appear nearer to the spine of the book than the other. Move your viewing eye until this reflection lines up with the edge of the book, as shown in Fig. 30. The other reflection will then appear beyond the edge of the book. You can measure how far beyond by bringing with you a piece of paper which carries a scale marked in inches for just this purpose.*

A simple proof indicates that this distance is the same as the distance between the first and second reflecting surfaces.

*If you have no scale, mark a piece of paper with the distance between the reflected images and measure it later.

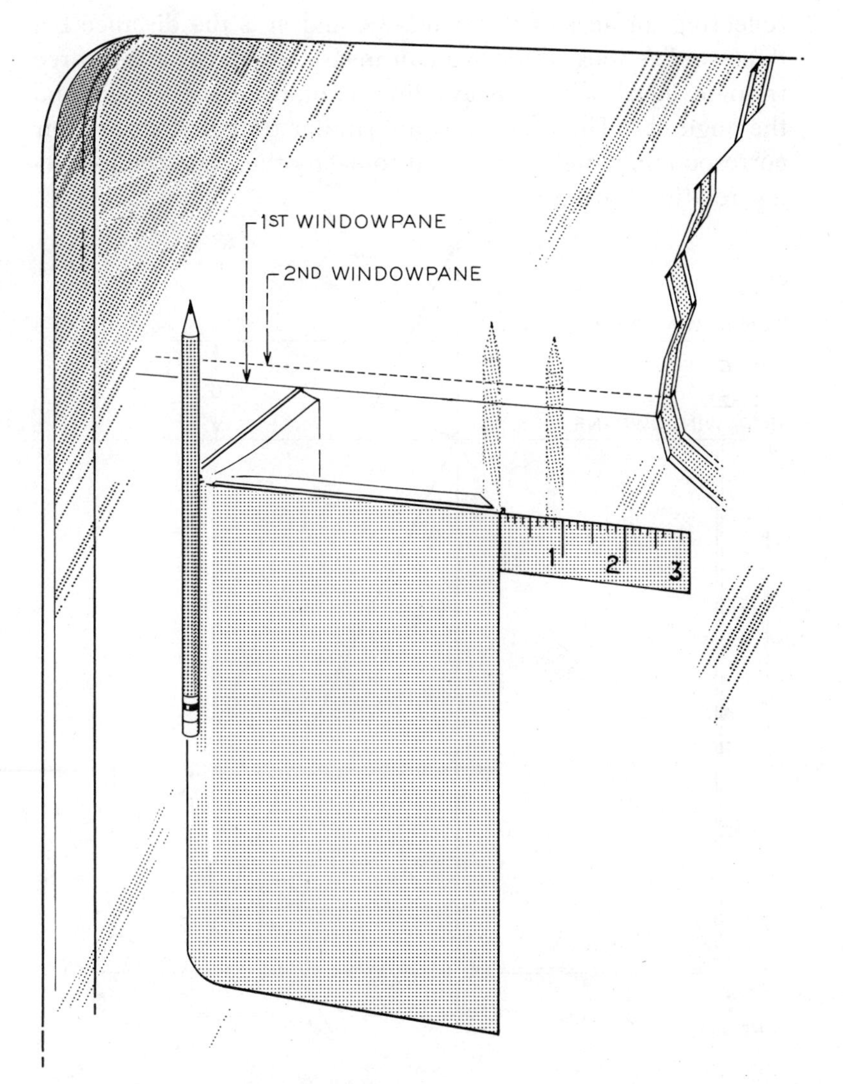

Figure 30. Double images reflected from two windowpanes

Referring to Figure 31, we see that d is the distance between reflecting surfaces of the windows, and m is the distance between reflections which we can measure. Each of the three triangles marked with heavy lines contains a right angle and the angle A. Therefore they are similar triangles, since their corresponding angles are equal to each other, and the following relationships are true.

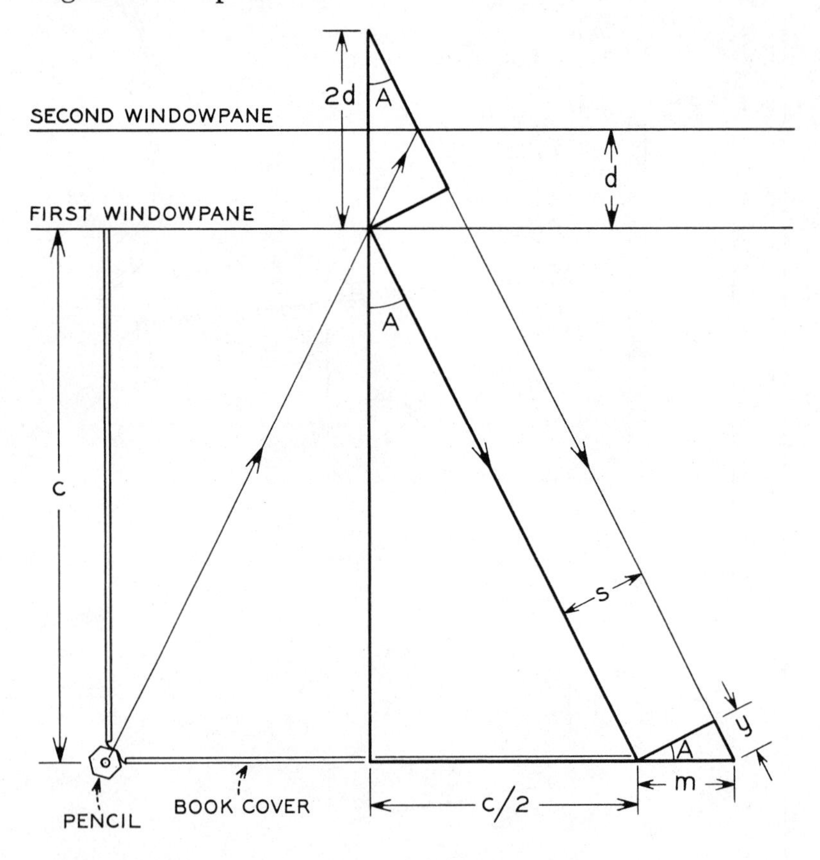

Figure 31. Diagram for analysis of double reflection

$$\frac{2d}{s} = \frac{m}{y}$$

$$\frac{2d}{m} = \frac{s}{y}$$

$$\frac{s}{y} = \frac{c}{c/2} = 2$$

$$\frac{2d}{m} = 2$$

$$d = m$$

In order for the observer to look along the two parallel lines (separated by distance *S*) simultaneously, he would have to be an infinite distance away from the windows, which is not possible, even in present-day airplanes. The visible reflections from the two windows travel in directions which converge to a point at the eye. Therefore the distance between the two images of the pencil will always be a distance (m') less than m in Figure 31 and will depend on the distance between the eye and the measuring scale. When the eye is 8 inches (a hand span) from the measuring scale, the distance between the reflecting surfaces of the windows (m) is approximately equal to $4m' - \frac{1}{2}$ in inches.

Some airplanes have more than two windowpanes. You then have your choice between ignoring all except the first two reflections (the two nearest to the spine of the book) or dealing with the more complicated relations of double reflections. Three windowpanes result in at least five images! (See Figure 32.)

If you were on the sunny side of the plane while you were making these measurements, you may have noticed the scratches on the outside of the window which were reflecting the sun brightly. They were not randomly oriented.

If the sun's rays are meeting the windowpane at a high angle (not far from perpendicular to the surface), the bright scratches may outline a set of concentric circles, but if the sun's rays meet the surface at a lower angle (farther from the perpendicular), the reflecting scratches will appear as a set of parallel straight scratches. Actually the scratches are quite randomly oriented on the windowpane, but the only scratches that shine brightly at any particular moment are those whose surfaces are at the proper angle at that moment for reflection of the sun's rays to your eye. When you are looking toward the sun, its rays seem to diverge toward you just as parallel railway tracks seem to diverge toward the observer. The windowpane traces of those reflecting surfaces that are perpendicular to the radial rays form the concentric circles.

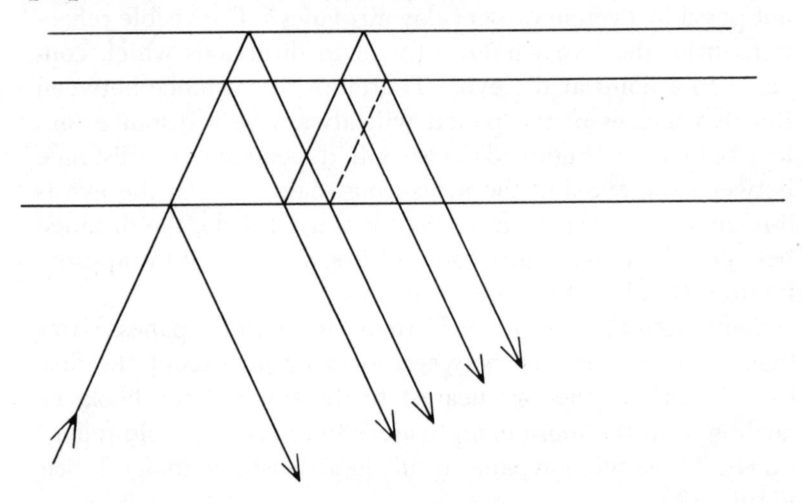

Figure 32. Five images from three windowpanes

When the reflecting scratches appear as parallel straight lines, they will be more nearly vertical when the sun is low and more nearly horizontal when the sun is high. However we cannot determine the angle of the sun's elevation by measuring the angle that the bright scratches make with the vertical since this does not tell us fully the orientation of the tiny reflecting facets in the scratches. We only know the direction of the scratches along the windowpane.

In spite of not being able to use the steepness of the reflecting scratches to measure precisely the angle of elevation of the sun, it may be amusing to note that they change direction more rapidly when one is flying east than when one is flying west.

9. *Frames of Reference*

AT EVERY MOMENT of our lives we are somewhere in space and somewhere in time. The ratio between a change of our position in space and the time that it takes it to happen is our speed. That sounds perfectly simple and straightforward. But suppose you are on a jet plane from New York to San Francisco and someone asks you what time it is. You may answer, "Well, by my watch it's eleven o'clock, but that's New York time. We left New York at nine and we're due in San Francisco at 11:55 their time, three hours earlier than New York time, so we're due there in three hours and fifty-five minutes and we've been in the air two hours. If we're $\frac{2}{6}$ or $\frac{1}{3}$ of the way, we're probably in the Central Time belt which is one hour earlier than New York time. In answer to your question, sir, it's ten o'clock". Clock time depends on your frame of reference.

When you get to San Francisco your watch will read 2:55 and you will reset it to 11:55. If you stay in San Francisco the rest of your life, will your life be three hours longer? Your next birthday will come three hours later, but that's just because it takes the sunrise three hours longer to reach San Francisco after reaching New York, as the earth turns.

If you take off from New York at 4:30 P.M. on a spring evening when the sun is due to set at six you will be taking off only an hour and a half before sunset. The sun is due to set in San Francisco at six o'clock San Francisco time. You will arrive there at 7:30 San Francisco time, only an hour and a half after sunset. If there are clouds in the sky reflecting the sunlight you may enjoy the spectacle of a sunset that lasts nearly six hours.

On a later plane you might watch the sun go down as you taxi to the runway for a takeoff. Then it will rise again for you when you get up away from the earth and set again a second time later.

Suppose you could fly around the earth in twenty-four hours and could continue to do so, without stopping for fuel or supplies. It would always be the same time where you were — noon, for example — but once in every twenty-four hours the date would change. Your motion with respect to the sun would be simpler than it is at present.

Our present system of naming the hours is considered by some to be out of keeping with our present way of life. It belongs to a time when people didn't move around so much and each place could have its local time without inconvenience. Suppose we called it twelve o'clock everywhere on the earth at the same time. This would greatly simplify all long-distance communication and travel arrangements. How long would it take us to get used to the idea that when people are having lunch in London at twelve o'clock, people in Rio de Janeiro are opening their business offices for the day at twelve o'clock, and people in New York are having breakfast at twelve o'clock?

Does our location in space present fewer problems? Suppose, for example, that we are 30,000 feet in the air above the Mississippi River at St. Louis, Missouri. This is a long way up. If we dropped a kitchen sink from this elevation it would take it $\sqrt{\frac{30{,}000}{16}}$ seconds to reach the ground, neglecting the retarding effect of air resistance. This is about 43 seconds or nearly three quarters of a minute. If you mentally drop the sink and wait three quarters of a minute while it hurtles down through space, the 30,000 feet will seem a very long distance.

We could try to indicate our elevation above the earth's surface with a model. Suppose we use an orange for the earth, a good big one, about three inches in diameter. This is about 7.9 centimeters and the diameter of the earth is about 7.9 thousand miles so this gives us a model on a scale of 1 cm = 1000 miles. On this scale, 30,000 feet, which is 5.68 miles, will be about 0.006 centimeter since it is 0.006 of 1000 miles. This is 0.06 millimeter which is about the diameter of a thin human hair. Hold a thin human hair close against the surface of an orange to appreciate how very close to the surface of the earth we are when we fly at 30,000 feet.

The rate of changing our position in space is our speed. But this is the most difficult concept of the three. As we have seen, the plane's speed with respect to the air is usually not the same as its speed with respect to the land. When it is standing at the airport, is its speed zero? Yes, with respect to the spot it's standing on, but that spot is rotating around the earth's axis moving through space at thousands of miles per hour. Is it moving relative to other spots on the surface of the earth? At first you would say no, since it gets no nearer nor farther from them. But suppose you were suspended in air at a point over the North Pole, looking toward the United States. As the earth turned, the United States would move from your right to your left, while behind your back Siberia would move from left to right. Would you say these two spots were not moving, relative to each other?

Is the North Pole, then, stationary? Not with respect to the sun as the earth revolves around it in its yearly orbit. Is the sun, then, our final frame of reference for motion? Is it stationary? The word has no meaning unless you define your frame of reference. It is impossible to speak of the state of motion of a body without comparing its position to that of at least one other body. Is the sun stationary with respect to what? In the great vast space of the universe bodies are moving with respect to each other. Choose one to use as your reference and speak of the motion of the others relative to that. The choice is arbitrary; there are no fixed walls to the room of the universe.

During the day the sun is always shining in the stratosphere. If you are on the side of the plane away from the sun, you can see the shadow of the body of the plane on the wing. If the plane is flying very steadily the shadow will be still. Note its position precisely and see how straight a course the pilot steers. Its position will change very slowly, of course, as the time of the day changes, but you may also see it change suddenly a small amount to a new position which will then be steadily maintained as the pilot resets his course from one straight-line air lane to another.

Today's airports frequently get so busy that planes have to wait their turn to land. Since they cannot hang motionless in the sky, the control tower assigns to each plane a race-track shaped "holding pattern" in which the plane travels lazily. The wide semicircular ends do not require detectable banking of the plane, and you may first become aware of your reversing route by noticing that you are alternately in the sun and in the shade.

Where a line from the sun through your head meets a cloud, you may see the shadow of the plane, a bright spot, the front end of a contrail shadow, or the center of a glory, as described in Chapter 5. This line is a reference direction in space which shifts only slowly with the passage of time. If the spot on the cloud shifts toward the front or the back of the plane in a short time, you know that the plane has changed its course.

At night, above the clouds, the stars are always shining. You can use them, just as you did the plane's shadow, to check the plane's course, especially if you are familiar with the constellations so that you can note the orientation of the wing tip relative to a known group of stars with confidence.

The stars are not a convenient frame of reference in our daily lives. When we are on the surface of the earth, we speak of motion relative to the earth's surface. When we see the trees go past the car windows we know we are moving. In the air, the far clouds appear almost stationary, but we see the near clouds fly past the window. Which way do the far clouds appear to move? If you fix your attention on a cloud in the middle distance, the clouds beyond it will appear to move toward the front of the plane while the nearer clouds will appear to move toward the back of the plane. If you shift your attention to a farther cloud, then it will become the stationary cloud. This is simply the phenomenon of *parallax,* which the dictionary defines as "the apparent displacement (or the difference in apparent direction) of an object as seen from two different points . . .". It is a phenomenon that can result in getting different readings from a meter, according to the direction from which one views the meter needle against the scale behind it. Close your left eye and hold your two index fingers up before your face, one finger lined up behind the other. Now open your left eye and close your right eye. Did the near finger jump to the right or did the far finger jump to the left? The answer will depend on whether your attention was fixed on the near one or the far one.

A stratocumulus cloud layer like that in the foreground of Plate XXIX may have gaps in it through which the airplane passenger can look down into the ground-floored space below, as one might look into a room through a skylight. As shown in Figure 33, such a gap will first line up with objects on the ground far ahead of the plane (position 1), then with objects directly below the plane (position 2), then with objects behind the plane (position 3). In other words, the hole, with its surrounding clouds, appears to move toward the tail of the plane *relative to the ground below.* If you raise your eyes and look out over the "cumulus meadows", as Guy Murchie calls them,

Figure 33. The appearance of relative motion, due to parallax

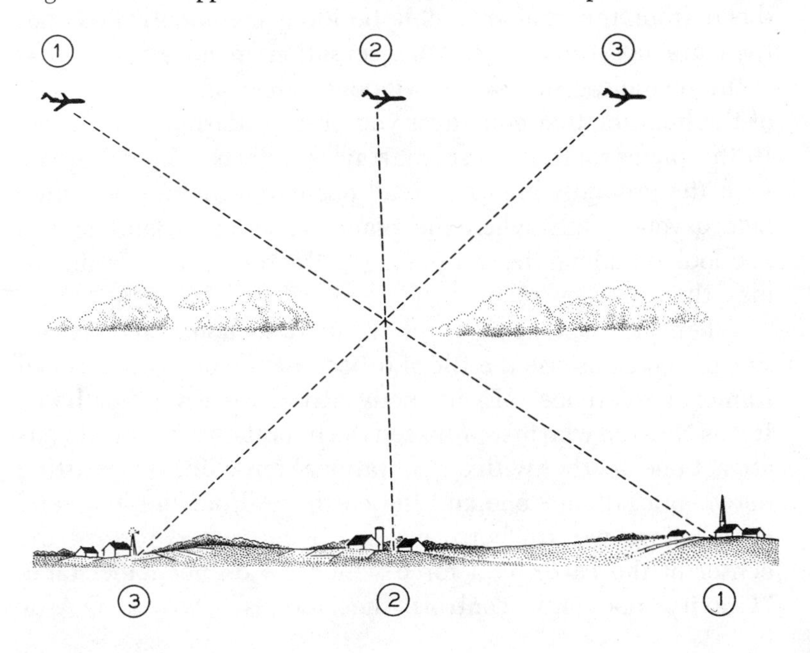

where the gaps are hidden because of the angle at which you are viewing them, the clouds will appear to come to an abrupt halt because the parallax effect is missing. Look down again and instantly they speed rearward over the ground.

You can sometimes use parallax to discover that there is more than one level of clouds below you; the upper clouds speed rearward relative to the lower clouds.

Little airplanes flying below you always appear to be moving toward the rear of your plane, relative to the ground. When they are flying in that direction, this makes them appear to be very swift; when they are flying in the same direction as your plane they seem very slow and may even appear to be backing up. When they are crossing the route of your plane, they move crabwise!

People who experience a panic fear of heights when looking down from the roof of a tall building commonly find that they are not troubled by this sensation in an airplane. It is the dwindling size of the window frames all down the side of the building that convinces you of your alarming elevation. In the plane there is no such array of objects connecting you with the ground which is spread out like a picture or a map before you. Only when the plane banks for a landing and you look out along the steep wing to the tops of near buildings may the sensation recur.

When an airplane is climbing or descending at constant speed, you can sense the tilt of it because of your gravitational frame of reference. Every body attracts every other body. It was Newton who first proposed the hypothesis that all objects attract one another with a gravitational force like that existing between a falling stone and the earth. All our life long, and even before we are born, we are being pulled toward the center of the earth by a force which we do not understand. "Gravity not only controls the actions", writes D'Arcy

Thompson*, "but also influences the forms of all save the least of organisms. The tree under its burden of leaves or fruit has changed its every curve and outline since its boughs were bare, and a mantle of snow will alter its configuration again. Sagging wrinkles, hanging breasts and many another sign of age are part of gravitation's slow relentless handiwork."

The angle by which the plane noses down or noses up is called the *angle of pitch.* Can we measure the angle of pitch by using the gravitational frame of reference?

First we need to know the direction of the earth's gravitational force relative to the airplane when the plane is flying level. To find this, stand this book upright on the left arm of your chair and suspend a key chain or weighted string from the corner as shown in Figure 34. (A string could be fastened into the book as described on page 77). The weight and the earth attract each other. The weighted string points toward the center of the earth. If the chair arm is not horizontal, the position of the string on the bookback goniometer scale will not be zero, but it will be the reference reading (for level flight) to which later readings will be compared. The angle read is the angle of inclination of the chair arm when the plane is flying level. Used in this way, the goniometer could be called a *clinometer.*

As the nose of the airplane tilts downward just before landing, the angle reading will change. If the plane descends at constant speed, the difference between the new reading and your reference reading will be the angle of pitch of the airplane. If the plane accelerates or decelerates, then the inertia of the weight on the string will cause it to "hang back" or "hang forward" and this effect will modify your pitch

* *On Growth and Form* (Cambridge, England, Cambridge University Press, 2nd ed., 1952), p. 51.

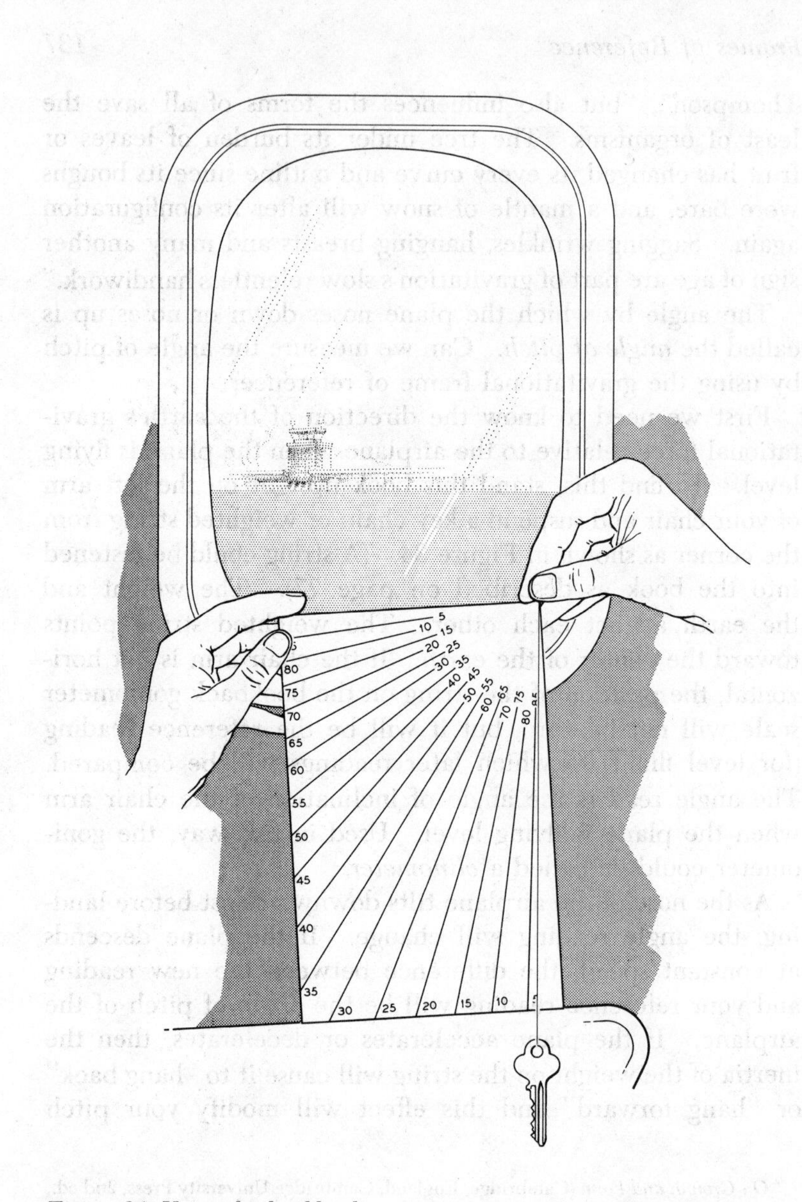

Figure 34. Using the bookback goniometer as a clinometer

measurement. You will not be able to tell how much of the angle is due to pitch and how much is due to the inertia of the weight during change of speed. (Further measurements related to landing are described in Chapter 14.)

Does the body of the plane on the ground have the same orientation as it does in the air or is its nose higher as it stands at the airport? We can determine this by comparing the reference reading of the clinometer on the ground with the reference reading when the plane is flying level. The reference reading on the ground can be useful to us. It enables us to use the instrument as an *accelerometer,* to measure the acceleration of the plane independent of its pitch.

Take a reference reading when the plane is on the ground and standing still. With the instrument in the same position (the book being on the left arm of the chair as in Figure 34 and parallel to the windows of the plane, with the chain or weighted string hanging from the corner), read the position of the string as the plane speeds up (accelerates) along the runway before taking off. How can we use this angle to determine the acceleration of the plane?

The force of gravitational attraction between any body and the earth causes that body to accelerate as it falls toward the center of the earth unless it is prevented from doing so. The propelling action of the jet motors or of the propellors causes the plane to accelerate forward and the weight on the string lags because of its inertia. The angle between your reference reading and the position read during acceleration measures the ratio of these two accelerations to each other, as shown in Figure 35. The tangent of that angle is the acceleration of the plane, a, divided by the acceleration due to gravity, g, which is known to be 32 feet per second per second or 32 ft./sec.^{2} or about 22 miles per hour per second. Since we are interested in how much the plane's speed (in the familiar miles per hour) changes in each second of travel down the runway, it will be more meaningful to use the hybrid unit, mph/sec.

Figure 35. Using the bookback goniometer as an accelerometer

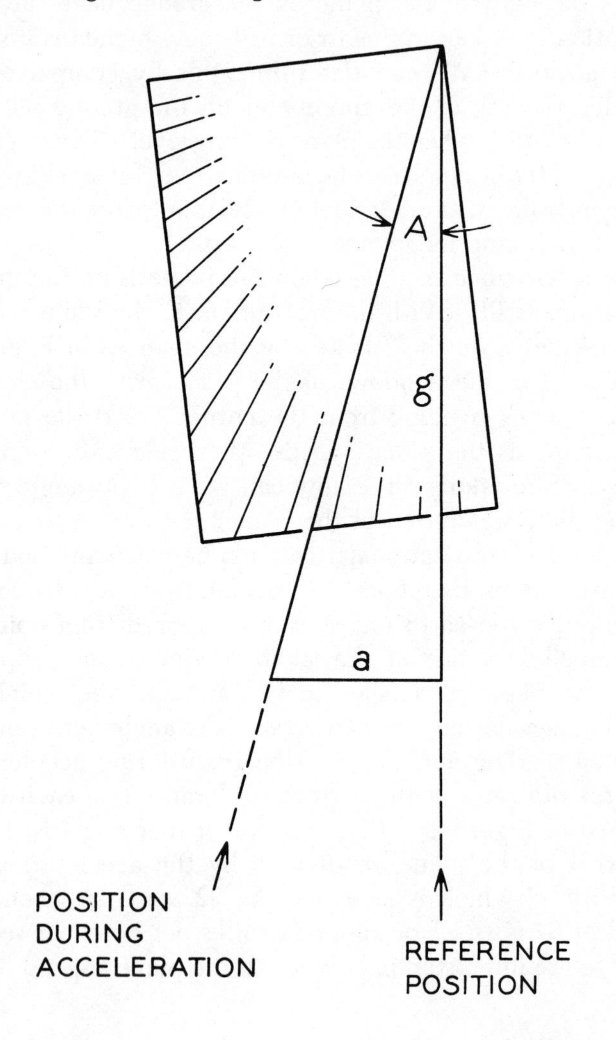

Suppose your angular-difference reading, A, is 16°

$$\tan A = \frac{a}{g} = \frac{\text{acceleration of the plane}}{\text{gravitational acceleration}}$$

$$a = g \tan A$$

$$a = 22 \tan 16° = 22 \times .287 \text{ (from Table 2)}$$

$$a = 6.3 \text{ mph/sec.}$$

If the plane continues to accelerate at this rate for fifteen seconds, at the end of that time it will be going 15 × 6.3 mph faster than it was at the beginning of that time, or 94.5 mph.

Are these figures probable? Could the plane take off at ninety-four miles per hour? How long does a plane continue to accelerate along a runway? Could you detect the effect when the front wheels leave the ground before the rear wheels do? These questions can be answered by measurements that you can make as the plane accelerates down the runway. When it takes off, the effect of the angle of pitch will be added to that of the acceleration and you will not be able to separate the two effects from each other.

Figure 36 is a graph of acceleration (a) in miles per hour per second versus the angle A, measured with the accelerometer. Note that the acceleration in these units is approximately equal to four-tenths of the numerical value, in degrees, of the measured angle, A. The graph is not a straight line and this relationship is only approximate, but you may wish to keep it in mind as you make the measurement.

When you are flying at moderate altitudes at night and there are no stars and you cannot see the ground, how do you know you are moving? If there is no change in the plane's velocity, no acceleration or deceleration upward or downward or sideward or in the direction of flight, you have no way of knowing. If you were to drop a ball in the airplane, it would not lag behind because it would leave your hand with the same forward velocity as that of the plane itself and there would be no force on it in a backward direction to slow it down. "Every body perseveres in its state of rest, or of uniform motion in a straight line, unless it is compelled to change that state by forces impressed thereon." This is Newton's first law of motion which we came upon in the first chapter of this book where we found that we sensed the change in

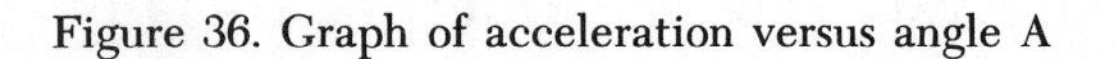
Figure 36. Graph of acceleration versus angle A

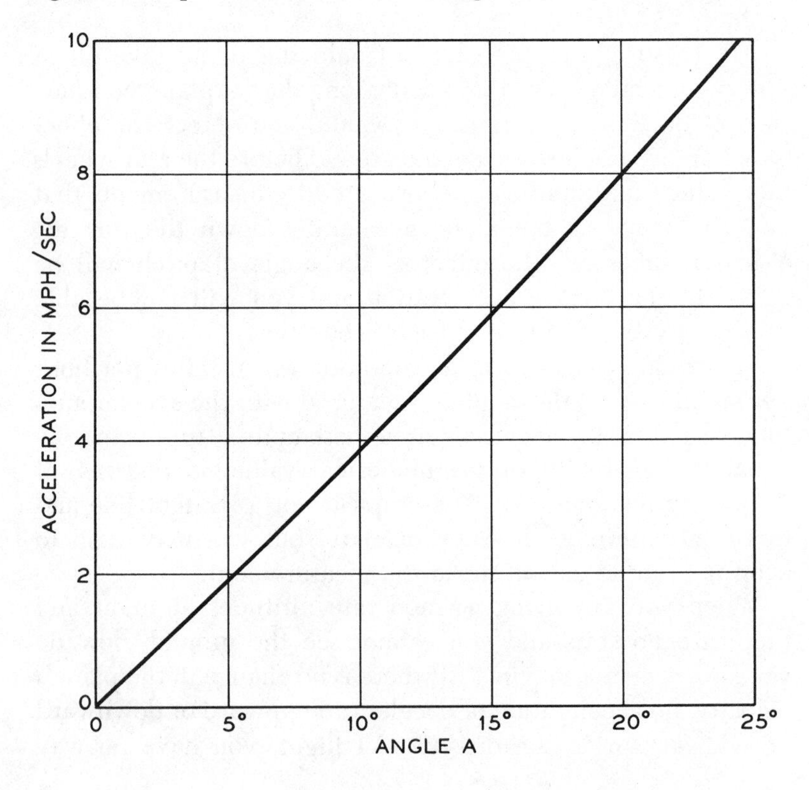

velocity of the plane because of the inertia of our bodies. Inertia gives us a frame of reference that stays with us when we cannot see the ground, the clouds, the sun or the stars. In the blackest night, when the airplane changes its velocity, we know it because of our inertial frame of reference.

The term *velocity*, as used by the physicist, refers not only to speed, but also to direction and it is used in this sense here. A change in direction of the airplane's course, even if the airplane maintains constant speed, is felt because of the inertia of your body. Forces must be exerted on it to change its direction of motion. If we could measure these forces accurately, we would know accurately what changes in the speed and direction of motion of the plane had taken place. By knowing the duration of the change and the length of time between changes, we would be able to plot its whole course, provided we knew where it started.

Just such an *inertial navigation system* is soon to be in use in commercial airplanes. Produced by the Sperry Gyroscope Company in 1966 for Pan American World Airways, it consists of gyroscopes as the inert masses, very sensitive accelerometers which measure the forces on the inert masses that occur whenever there is a change of velocity, a clock, and a digital computer to keep track of everything. The initial latitude and longitude of the plane are recorded in the computer, given in degrees, minutes and tenths of minutes. From that point on, the accelerometers measure all changes relative to the inertial frame of reference (which is maintained by the gyroscopes), the clock measures the time during changes and between changes. From these data the computer calculates continuously the latitude and longitude of the airplane and displays this information to the pilot and copilot.

The accelerometers which are measuring changes of velocity in the horizontal plane by the very small forces produced must not be allowed to tilt at all, for then they would include some of the gravitational force in their measurements. The gyroscopes must maintain a "stable platform" perpendicular to the gravitational attraction of the earth and with a fixed orientation around this direction as an axis. This establishes the inertial frame of reference, relative to which the airplane's movements are measured.

An interesting problem is posed by the fact that the takeoff point and destination of the plane are constantly moving as the earth moves. Suppose a plane starts northward from the equator to reach a destination due north from its starting point. Every point on the equator is moving eastward at more than 1000 miles per hour as the earth's rotation causes it to move along a 25,000-mile circle in 24 hours. So the airplane is moving eastward with the earth at this velocity when it starts. Farther north, points on the surface of the earth move along smaller circles during the 24-hour rotation, so if the plane does not change its crabwise eastward velocity it will arrive east of its intended destination. The farther north it goes, the greater the correction required. From the pilot's point of view, he must keep turning a little toward the west in order to travel due north. The plane follows a curved path relative to the inertial frame of reference in order to follow a straight path relative to the surface of the earth. The inertial navigation system stores the necessary information in the computer and it is taken into account when the computer reports the latitude and longitude.

Each of us carries within his own head a device which enables him to detect the motion of the head with respect to an inertial frame of reference. Part of it enables us to detect rotational acceleration and part of it enables us to detect changes in translation, displacement without rotation. Both parts belong to the structure of the ear.

The so-called *semicircular canals* are three interconnected tubes (Fig. 37) of membranous material supported by a bony structure of the same shape. There is a fluid inside the membranous tubes as well as between them and the bony walls that surround them. When the head is turned the fluid lags behind, due to inertia. The motion of the tubes relative to the fluid is detected by the nerves distributed along the walls of the tubes. The canals lie approximately in three mutually perpendicular planes so that each is well suited to detect changes in rotation around a particular axis. Figure 37 shows, for example, that the lateral canal is best suited for detecting changes in rotation around a vertical axis (an axis parallel to the long direction of the page) since, for such rotation, the motion of that tube along its own length is a maximum, relative to the inertial fluid. The other canals detect changes

Figure 37. The semicircular canals of the ear

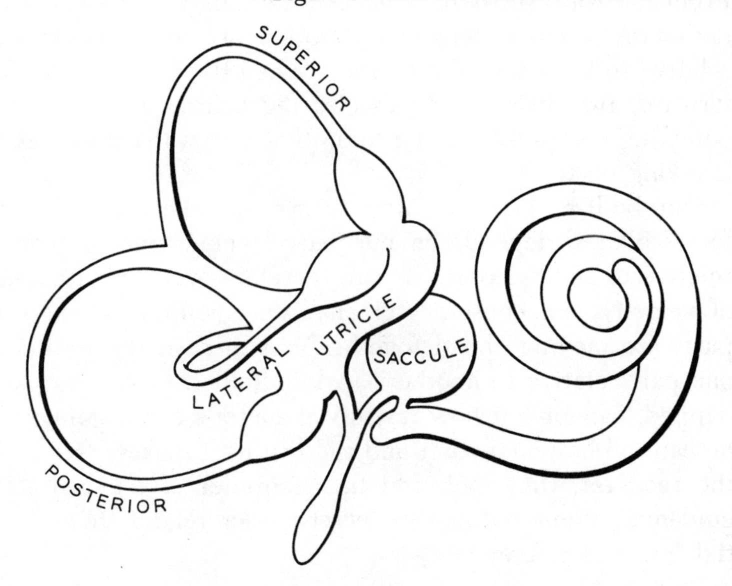

in rotation around two horizontal axes and all rotations can be resolved into these three components of rotational motion.

A purely translatory acceleration, as when the whole head starts to move forward without turning, cannot be detected by the semicircular canals. Although the fluid in each limb of the curved tube tends to lag when the tube starts to move forward, the motions in the two limbs oppose each other, one being clockwise and the other counterclockwise. The fluid is so incompressible that it does not squeeze down to the back end of the canal when the head moves forward.

To detect translatory acceleration we need objects in a container, like people in an airplane, that will lag behind when the container accelerates forward. The little sac or saccule of the ear (Fig. 37) contains a gelatinous substance in which are suspended some solid particles of calcium carbonate which perform just this function. The utricle is similarly equipped. From the walls of these pouches tiny hairs interspersed with unsheathed nerves detect the motion of the solid particles relative to the walls of the sacs. When the head accelerates forward, the little solid particles lag behind a bit and the detecting nerves inform the brain that a forward acceleration is taking place.

Our built-in inertial guidance system, like that designed for airplanes, depends on one part lagging, due to inertia, relative to another part. It can therefore only detect change of velocity, not constant velocity, since without change all parts are moving ahead together and there is no motion of one part relative to another. A blindfolded man, being kidnapped, cannot tell how fast his abductor's car is going, but he can tell how many left and right turns it makes. He lacks the recorder, the clock and the computer of the complete guidance system, but he can detect motion relative to an inertial frame of reference.

10. Rock Structures

"AERIAL PHOTOGRAPHS today are widely used to obtain both qualitative and quantitative geologic information . . ." So begins Geological Survey Professional Paper 373, *Aerial Photographs in Geologic Interpretation and Mapping* by Richard G. Ray, published by the United States Government Printing Office in Washington in 1960, and a big bargain at $2.50 with nearly 100 aerial photographs and interpretative captions.

To see why today's geologists take to the air we need to review briefly how rocks are formed. When molten rock deep within the crust of the earth becomes cool enough it crystallizes, just as molten ice from the water faucet crystallizes when it is cooled in the freezer. The geologist calls the molten rock *magma,* which is an old word for porridge.

The solid crystalline rock formed from magma is called *igneous rock* and there are many different kinds with different colors and textures according to the composition of the magma and its rate of cooling. If it pours out on the surface it is called *lava.* Lava commonly cools so quickly that the atoms don't have time to get into the orderly arrangements that characterize those solids that we call crystals, so *volcanic glass* is formed. This is one of the few naturally occurring noncrystalline solids. If the magma cools more slowly, below the surface, crystals grow, neighboring crystals being separated from each other by irregular boundaries, as one can see in so many examples of polished igneous rocks used for decorative trim at the street level of our large buildings. All crystals that form naturally (as contrasted with those grown in the laboratory) are called *minerals.*

Since the land is constantly being worn away, mostly by the action of running water carrying sediments, but also, to a much lesser extent, by the action of wind and ice, these igneous rocks ultimately become exposed at the surface. Frost action, plant action and chemical attack (oxidation, hydration and the forming of carbonates by reaction with the atmosphere) break them down into particles which can be carried by water. Even the general movement of rainwater down a slope where no stream valley exists results in the transfer of particles downward. Swiftly moving streams can carry large particles along with smaller ones. If the velocity of the water decreases, as it does when it enters a lake, it can no longer carry its load and will deposit it, dropping the coarser particles first and the finer particles as it slows down still more. The finest silt will settle slowly to the bottom. Sediments deposited in this way in lakes and bays and seas over millions of years have become cemented into *sedimentary rocks,* whose layered structure and sorted grain-size attest to their water-borne origin. Both sedimentary and igneous rocks may become modified by heat and pressure and the great variety of solutions that permeate the earth's crust. Rocks thus *changed* in *form* are called *metamorphic rocks.*

A skilled geologist, trained in the interpretation of aerial photographs, can distinguish each of these types and their structural relations from favorable photographs, but even the most unskilled amateur can spot some of them from the air. In Plate XVII the rejuvenated Little Colorado River is cutting down through rocks that can be seen as horizontal ledges along the side of the river. From the top of the ledges nearest to the river, a broad gentle slope extends back on either side to more ledges, just like those near the river, but higher and farther back. The broad area must have been the valley floor at an earlier time when the river in its various meandering positions cut against the front surfaces of the higher ledges as it is now cutting the lower ones.

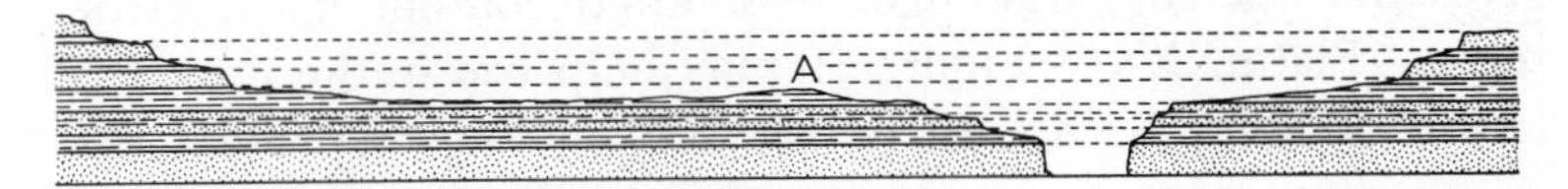

Figure 38. Diagram suggesting geologic structure near Cameron, Arizona (The stippled layers are those more resistant to erosion)

If we were to cut a vertical slice through the land, from left to right across Plate XVII, about three-quarters of the way from the bottom of the photograph (passing through the Little Colorado River where it shows a small white triangle of reflected light), the structure of the rocks we could then see would probably look much like the diagram in Figure 38. Where there are gently sloping eroded surfaces, the rock is less resistant to erosion by the river. Cliffs mark those rocks that were more difficult to erode. The matching layers on opposite sides of the river and the continuity of ledges along the river are evidence of the horizontality of the rock layers. Clearly they were continuous before they were worn away by the river. These are sedimentary rocks, formed from sediments that settled in waters that covered this region long before the existence of the present rivers.

If you could inspect these rocks at close range you would

find them made up of layers of particles, sorted by the carrying power of the water so that any one layer consists of particles of uniform size. From the air this layered sorting shows up as slight differences in shade. On the small low hill just back from the inner gorge of the river on the left (A, Fig. 38), the photograph shows curving bands of light and dark rock where the edges of the horizontal beds crop out at the rounded surface of the low hill. Compare the similar curving of the horizontal crop strips on the low hill in Plate XXXVIII.

In the vicinity of Redford, Texas, near the Rio Grande River the pattern of the outcropping edges of the horizontal beds gives the airplane passenger a detailed contour map of the region (Plate XXX). Figure 39 may help you in the interpretation of Plate XXX. In it, the broken lines represent the

Plate XXX. The Rio Grande near Redford, Texas

edges of selected horizontal sedimentary-rock layers in the photograph; solid lines represent stream valleys in the photograph and the double line represents the lonely road across the flat upland between the valleys. At the bottom of Plate XXX, in the center, two short streams flow out of the photographed area. Follow the right-hand valley upward to its dark pointed end and see how the band of the rock outcrop beyond it bends sharply away from this point where the headwaters of the stream have eaten back the land. The headwaters of the little stream on the left have three well-defined branches, tributaries eating their way into the flat divide that separates these valleys from those at the extreme left.

Not all sedimentary rocks occur in horizontal beds. The slow inexorable motions of the earth's crust, whose causes

Figure 39. Sketch of Plate XXX (Double line represents a road; single lines, streams; broken lines, edges of sedimentary rock layers)

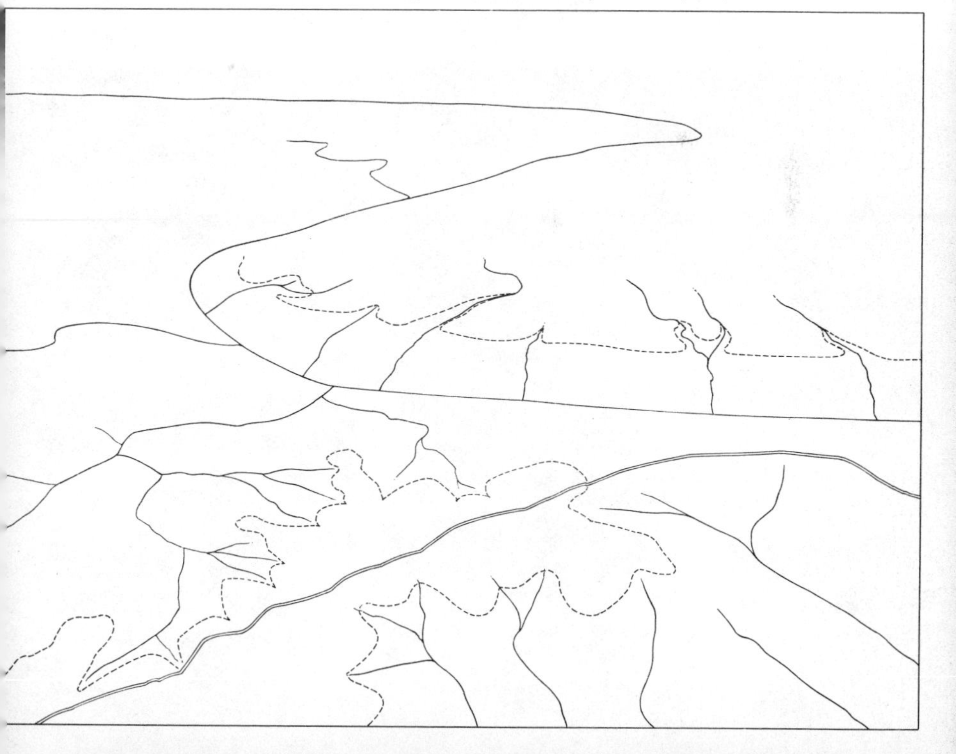

are even now not well understood, have resulted in folding and cracking of the originally horizontal layers after they had become "rigid" rocks. Rocks that will break if struck suddenly with a hammer can be folded without breaking by forces applied over long periods of time, just as a glacier will slowly flow down a valley. If you try to bend a candle in your hands you will break it, but if you lay the candle on a shelf, with nearly half of it extending beyond the edge of the shelf and leave it for a few weeks, it will bend under its own slight weight.

Folding of rocks, raising and lowering of the land relative to sea level, metamorphism of rocks — these changes are going on today beneath our very feet without our being aware of them except when a slight slipping occurs along some break in the rocks. Only then do we sense an earthquake.

When folded sedimentary rocks are eroded, they may show

Plate XXXI. Pitching anticline, eroded

clearly to the airplane passenger the nature of their structure. In Plate XXXI, for example, the differential erosion of layers of different resistance shows us that the layers were deformed into a fold which was higher in the middle, a fold which the geologist calls an *anticline*. A fold that is lower in the middle is called a *syncline*.

If we again made a vertical cut down into the earth as with a giant knife, slicing from left to right just above the center of the photograph, we would have a cross section of the structure of the folded layers that would probably look rather like Figure 40, in which, however, the vertical scale is exaggerated. We can easily imagine how each folded layer must have been continuous before a portion of it was removed by erosion. The missing part of the structure is shown by dashed lines in Figure 40.

The whole great fold pitches downward away from us in

Figure 40. Diagram suggesting geologic structure of anticline shown in Plate XXXI

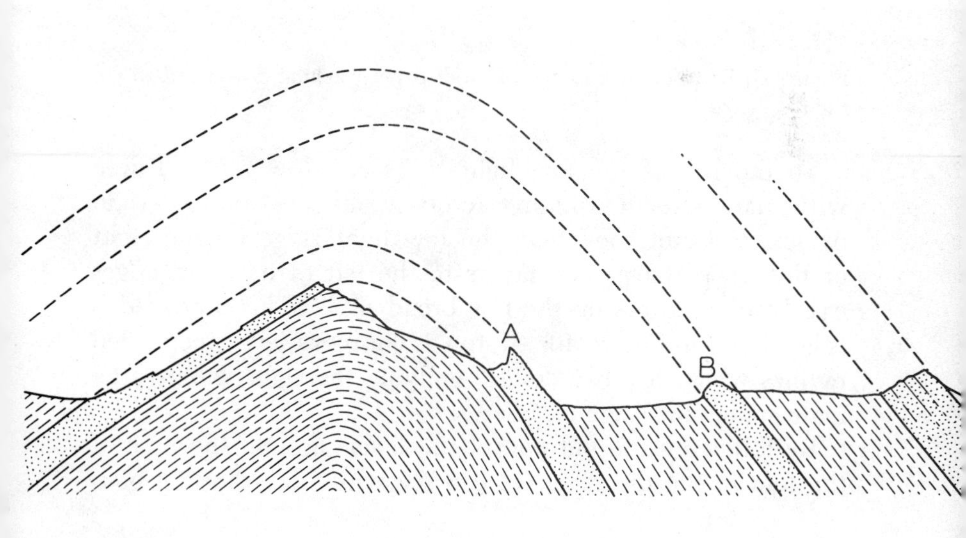

the photograph. The clearest evidence of this is the curved outcrop of the thin light-colored layer which turns sharply near the top of the photograph. (Figure 41 is a diagram of a folded piece of paper pitching downward to the right. The upper part of the paper has been cut off by scissors, leaving a curving edge, just as the folded rock layer was worn off by erosion, leaving a curving edge.) In Plate XXXI the narrow light-colored rock layer forms a low ridge (B in Fig. 40). Streams have eaten through it in places. At the right of the picture the sediments that choke these streams hide the light-colored layer where they cross it.

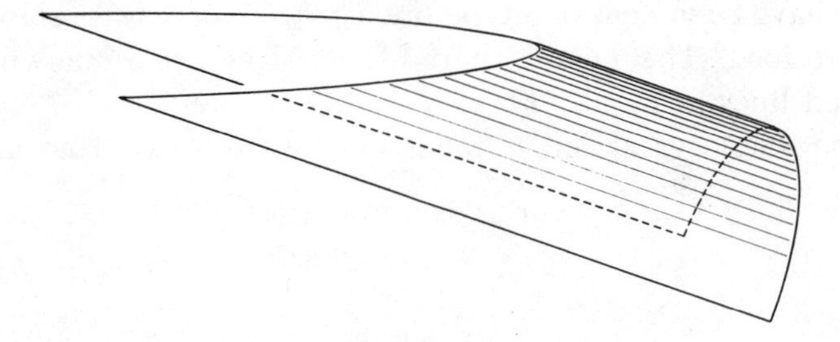

Figure 41. Diagram of a folded piece of paper pitching downward to the right

To the left of the low light ridge is a lower broad area with many streams running across it. It must be underlain by less resistant rock than the low light ridge on the right or the steep sharp rock layers to the left of it whose edges cast black shadows on the low broad area. This sharp ridge is labeled A in Figure 40. It, too, is cut by valleys, steep-sided ravines whose headwaters are gullying the right flank of the main mountain ridge.

By following the outcrop of the layers around the nose of the fold in the distance, we can see which left-dipping beds on the left are the continuation of which right-dipping beds on the right. It is easy to spot the narrow, light-colored bed B in the upper left corner and the broad unresistant layer inside (and therefore below) it, but it is harder to recognize the left-dipping layers of bed A because they do not stand up as a ridge on the left side of the mountain, but parallel the slope of the mountain at a somewhat less steep angle, overlapping, one on the other, in slabby sloping steps (Fig. 40).

Perhaps this more resistant layer is now protecting the left side of the mountain from being eroded away as badly as the right side has been. There is more vegetation on the right side. Perhaps more rain falls there, or protection from the afternoon sun may result in less loss of water to the air.

Recall that these layers were originally laid down horizontally, each bed of sediments on top of the one previously deposited, that they were then cemented into rock. With these facts in mind it should be possible to tell which of the rock layers exposed in Plate XXXI is the oldest and which the youngest, the most recently formed.

Flying over less arid country, one seldom finds the structure of the rock so bared to view. In humid climates the processes of chemical weathering decompose the exposed rock and the decomposed rock supports vegetation which in turn decomposes to give humus to support more vegetation.

In the Appalachian Mountains, however, from eastern Pennsylvania to Alabama, even the thick cover of trees and shrubs cannot hide the curving ridges of the pitching anticlines and synclines. The gentler slopes parallel to the sloping layers of the beds and the cliff-like slopes marking the eroded edges of the beds are best detected when the sun is low.

It is difficult to recognize the relief, the ups and downs of the surface of the earth, from an airplane. The irregularities appear to flatten out because our eyes are so close together. When an object is near you, your left and right eyes see it from different directions. Your depth perception depends on the comparison of the two different images. When you look at features thousands of feet away, your left and right eyes see them from almost the same direction and the two images are so nearly the same that you cannot derive depth information from them.

Sometimes there are hints about the relief. When a road zigzags back and forth, for instance, it is climbing a very steep hill. When a stream is white and the weather is not cold enough for ice nor dry enough for salt deposits, the whiteness must be due to rapids in a steeply falling stream. In a region of tilled fields or housing developments, a strip of woodland

Plate XXXII. A pitching anticline in the Appalachian Mountains near Harrisburg, Pennsylvania

may mark a slope too steep to plow or build on (Plate XXXII).

Water reliably gravitates to low spots. A little pond or even a damper area in a field will tell you that that spot is lower than its surroundings. Plate XXXIII shows an area in Kentucky peppered with small pits which show as little ponds and damp spots. Such *karst topography* results from solution of an underlying layer of limestone. Limestone is composed of calcium carbonate which is slightly soluble in rainwater and groundwaters. In humid climates it is not only worn away more easily than other rocks are but it is also eaten away by solution, so the land surface that it underlies has a moth-eaten appearance.

In mapping topography, the shape of the surface of the land, the aerial photographer takes two photographs of the

Plate XXXIII. Karst topography, indicating that the bedrock is limestone, near Flaherty, Kentucky

same area from two widely spaced positions and subsequently views one print with the left eye and the other with the right eye, as in the old-fashioned stereopticon viewer. Techniques have been developed for making accurate topographic maps from such stereoscopic pairs of photographs.

A flight over the island of Hawaii may afford a magnificent opportunity to see beds that have been built up by successive outpourings of lava so that they resemble sedimentary rock layers. Plate XXXIV shows Kilauea in the foreground and Mauna Kea in the background. The lighter rocks are older

Plate XXXIV. The crater of Kilauea in the foreground, Mauna Kea in the background, Hawaii.

lava that has been weathered by exposure to the atmosphere. The dark lava in the circular pool (now hard rock) is from eruptions that took place in 1952 and 1954. To the right of the dark pool a fissure opened in 1954 and let the dark lava flow out across the surface of older lava beds.

In times past the central region of the summit of Kilauea has slumped, breaking loose from the surrounding rocks. The cliffs surrounding the large central basin show where the break occurred.

One could go on through many pages, picking special features like the lava flows of Kilauea and the karst topography of Kentucky that can best be seen from the air. John Shelton has done this for over four hundred pages with nearly four hundred illustrations in a book called *Geology Illustrated.**
Many of the illustrations are aerial photographs which could well convince the student of geology that most of his classes should be conducted on an airplane.

*(San Francisco, W. H. Freeman and Company, 1966.)

11. *Vegetation*

WHAT GROWS where and why? Figure 42 shows the distribution of the major vegetation regions in the 48 contiguous states and part of Canada, and Figure 43 shows a more detailed subdivision into types of vegetation.

Plants need water and nutrient materials, the right amount of sunlight and suitable temperatures. The fact that different amounts of water, nutrients, sun and warmth are required by different types of plants results in the distribution shown on the maps. All this is the province of ecology, that branch of biology that deals with the relationships between organisms and their environment.

Within the broad vegetation-type regions shown in Figures 42 and 43, the airplane passenger can distinguish micro-regions that differ in vegetation because of local environmental differences. You may be able to discover the causes of those differences.

In semi-arid regions where the climate will just barely allow the growth of some plants, small differences in the amount of water may be a life-and-death matter. In some places the north slopes of hills are green and the south slopes barren, due to the difference in the angle at which the sun's rays fall on the surface. Everywhere north of the Tropic of Cancer the sun is always in the southern sky during the middle part of the day. This means that the sun's rays fall more nearly perpendicularly on the south-facing slope of a hill than they do on the north-facing slope. A beam of light must spread its energy over a greater distance on the north-facing slope than a beam of the same width falling on the south-facing slope. In this way the southern slope gets more heat per square inch from the sun than the northern slope does. In hot, semi-arid regions, therefore, the south-facing slopes may become too hot and dry to support vegetation that can survive on the north-facing slopes. In very cold regions, however,

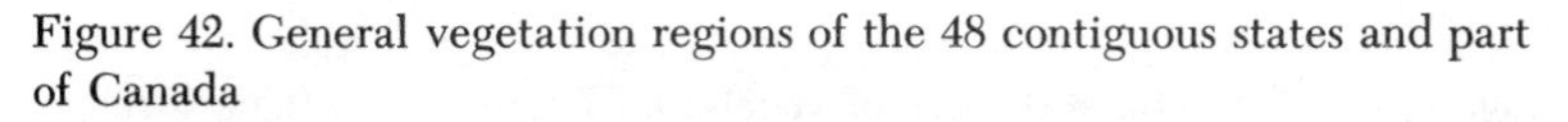

Figure 42. General vegetation regions of the 48 contiguous states and part of Canada

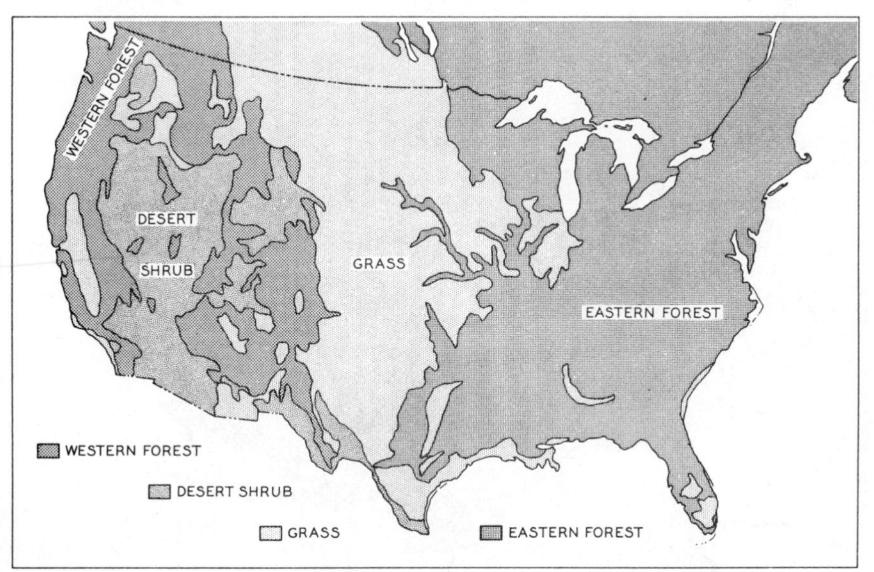

some plants can get through the winter on a southern slope that could not survive on the colder northern slope.

Plate XXXV shows three tributaries flowing westward to join a major stream in central Idaho. The tributaries are separated from each other by ridges. Immediately south of each stream the trees begin and continue up the north-facing slope of the ridge until, near the top, they find conditions too dry to support growth. On the south-facing slopes of the ridges vegetation is sparse or lacking altogether.

Where the prevailing winds are from the west and bring moisture, the western slopes of a mountain range may be covered with vegetation and the eastern slopes barren because they lie in the "rain shadow" of the mountains. The moisture-laden air goes up the west side of the hills, is cooled adiabatically (Chapter 6), and the moisture condenses as dew,

Figure 43. Characteristic types of vegetation in the 48 contiguous states and part of Canada. Right: Key to this map.

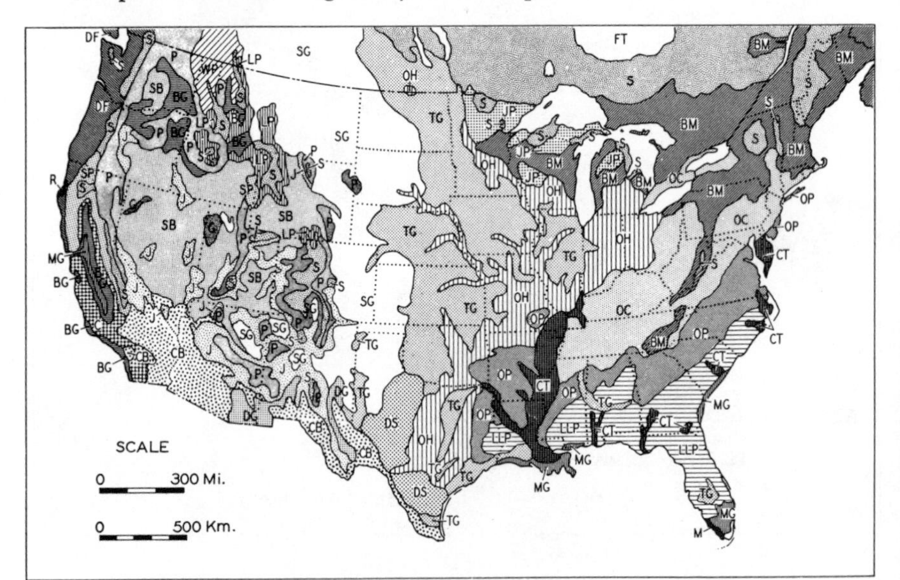

mist, or rain, making plant growth possible. As the air moves down the other side of the mountains it becomes warmer, its relative humidity decreasing rapidly. Not only does it seldom contribute moisture to that side of the mountains, but most of the time it takes up moisture.

Plate XXXVI shows an area just north of Sun Valley, Idaho, that illustrates this effect. Again, north is at the top of the picture. Near the left edge of the picture is a sharp ridge with bits of snow lingering along its top. Since the east side of the ridge is better lighted than the west side, the picture must have been taken in the morning. At the foot of this slope the Big Lost River of Idaho flows north. To the right of the river the west-facing slope is covered with trees which cast much longer shadows than the trees on the east-facing slope because of the direction from which the sun's rays are

EASTERN FOREST VEGETATION

FT Subarctic forest-tundra transition (Canada)
S Spruce-fir (Northern coniferous forest)
JP Jack, red, and white pines (Northeastern pine forest)
BM Birch-beech-maple hemlock (Northeastern hardwoods)
Oak forest (*Southern hardwood forest*):
- OC Chestnut-chestnut oak-yellow poplar
- OH Oak-hickory
- OP Oak-pine

CT Cypress-tupelo-red gum (River bottom forest)
LLP Longleaf-loblolly-slash pines (Southeastern pine forest)
M Mangrove (Subtropical forest)

WESTERN FOREST VEGETATION

S Spruce-fir (Northern coniferous forest)
Cedar-hemlock (*Northwestern coniferous forest*):
WP Western larch-western white pine
DF Pacific Douglas fir
R Redwood
Yellow pine-Douglas fir (*Western pine forest*)
- SP Yellow pine-sugar pine
- P Yellow pine-Douglas fir
- LP Lodgepole pine
- J Pinon-Juniper (Southwestern coniferous woodland)
- C Chaparral (Southwestern broad-leaved woodland)

DESERT SHRUB VEGETATION

SB Sagebrush (Northern desert shrub)
CB Creosote bush (Southern desert shrub)
G Greasewood (Salt desert shrub)

GRASS VEGETATION

TG Tall grass (Prairie grassland)
SG Short grass (Plains grassland)
DG Mesquite-grass (Desert grassland)
DS Mesquite and desert grass savanna (Desert savanna)
BG Bunch grass (Pacific grassland)
MG Marsh grass (Marsh grassland)
Alpine meadow (*Not shown*)

coming. Here, clearly, the western slopes are more favorable for the growth of trees.

In much of the southwestern United States trees grow only along the rivers where their roots can reach below the water table, the level of saturation of the soil. Many of these trees are cottonwoods. In such country the hills are usually sparsely dotted with small shrubs (Plates XXX, XXXI). In the north these are commonly sagebrush, in the south, creosote bush.

In farming country the banks of streams may be lined with trees for a different reason. The farmer can hardly plow right

Plate XXXV. Vegetation on north-facing slopes, central Idaho

Plate XXXVI. Vegetation on west-facing slopes near Sun Valley, Idaho

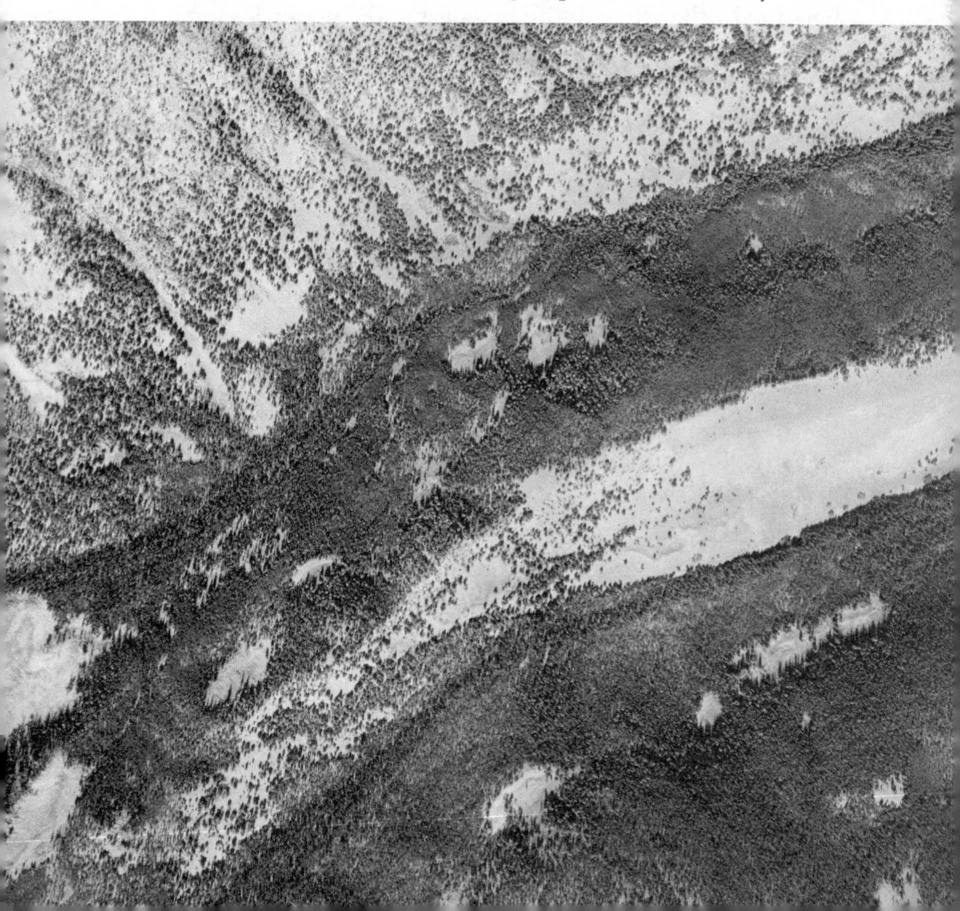

down to the edge of the water. The edge of the field stops where the slope of the river bank begins (Plate X).

It is not always so that there are more shrubs and trees where there is more water. In Plate IX most of the area is covered with salt-marsh grass. Shrubs and trees can only grow on the higher land that is better drained and not saturated with salt water.

Flying over wooded land, the average airplane passenger probably finds difficulty in distinguishing maple trees from oak trees, for example, and perhaps even in distinguishing the evergreens from the deciduous trees. As summer gives way to fall, some of the wooded areas take on significant color. In parts of the northeastern United States the wooded valleys become brilliantly colored because of the water-loving maples, while the hilltops turn brown because the oaks, which can tolerate a smaller supply of water, grow on the higher (and therefore drier) places and turn brown in the fall. The western cottonwoods, growing along the streams, turn bright yellow in the fall.

At this season, too, the cultivated fields change their appearance rapidly as the harvest is reaped. In some regions cornfields show an orderly array of dots which are the stalks of corn gathered neatly into shocks, leaning together in tepee-like bunches to dry out.

In winter, when snow is on the ground, it shows through the bare branches of the deciduous trees except for the oaks which cling to their brown leaves until spring. Brown patches of oak trees hide the snow as well as do the dense evergreen woods.

Rocks and roads and water and clouds may be very similar from one trip to another, whether in spring, summer, or fall. It is the ever-changing vegetation that gives variety to repeated trips over the same route.

12. Man-made Structures

PEOPLE build puddles of buildings at the intersections of major transportation lines. Fingers of population reach out along the roads from town, dwindling as the distance from town increases. One of the best ways to appreciate man's achievements in adapting to and controlling his environment is to get a bird's-eye view of them from the air. If man lived only where he could be comfortable without shelter, be fed from what grew naturally around him, and move from here to there without clearing a way, his distribution on the face of the earth would be very limited indeed.

In the big cities he has huddled together so closely that he has piled himself on top of himself to great thicknesses. From the air you can watch the average height of the buildings decrease as the distance from the center of the city increases: skyscrapers at the center, lower office buildings and apartment houses around the outside, ranch houses in the suburbs. Do big cities engender a cluster of smaller communities around them, identifiable from the air as separate units, or do they just dwindle off into the countryside in all directions? Are inter-town distances directly proportional to the sizes of towns: big towns far apart, middle-sized towns at intermediate distances, and small towns relatively close together? These are questions you can best answer from the air.

In an area where the houses are close together, you will sometimes see patches of woods. For some reason this patch is unused, probably unusable, for housing. The land is too steep, too rocky, or too wet. There may be other clues from the surroundings that will tell you which of these characteristics explains the presence of the wooded patch.

In country that is intensively farmed, wooded patches also mean areas that are too steep, too rocky, or too wet. Such patches may be seen in the farmed land shown in Plate XXXVII. The roughly triangular area near the right edge of the picture has probably been left unplowed because it

Plate XXXVII. Strip-cropping in northern New Jersey

is too wet, judging from the fact that the neighboring plowed strip shows dark wet spots. In this picture, the white strips are plowed land not yet covered by vegetation, the coarse-textured gray strips are young corn and the fine-textured gray strips are a cover crop such as alfalfa. This method of farming, which is known as *strip-cropping*, makes spectacular patterns for the airplane passenger. It is done to minimize removal of the soil by surface run-off water. Open crops, such as corn, that leave exposed soil between the plants, permit the removal of a little of this soil every time the rain falls. If the neighboring strip of soil bears a crop of tightly intergrown, surface-covering plants, such as alfalfa or hay, the motion of the water will be checked there and it will drop the bits of earth that it is carrying instead of removing them from the field altogether.

Plate XXXVIII. Contour strip-cropping on a low hill

Since the water trickles down the slope of land, the strip boundaries should be across the slope of the land, at right angles to the direction of the moving water. Since this will be the horizontal direction, the strip boundaries stay at uniform level, just as the lines on a contour map do. To do this, they must curve outward around the nose of a hill and inward (upstream) where they cross a valley. Where the plowed strips form closed loops, as in Plate XXXVIII, you are looking down on a rounded hilltop. In Plate XXXIX a little valley has been dammed to form a silt-catching pond. In the middle of the picture, on the right, the curving level strips indicate that there is a gentle slope upward toward the right. Farther back, toward the left, the curving strips indicate that the land

Plate XXXIX. Strip-cropping and a silt-catching pond

slopes upward toward the left. In between is the valley, with a small wooded area probably indicating land too wet to farm. The straight strips in the foreground of Plate XXXIX do not appear to conform to the contour plan of plowing, if indeed the land slopes toward the body of standing water. But perhaps the sandy bank near the water is steep and the field is in fact level. If the bank were steep it would appear to change in width as you flew over it.

In very flat country, you may sometime see a spirally plowed field (Plate XL). By starting at the center and plowing on an ever-increasing radius of curvature of the furrow, the farmer shifts the soil a bit toward the center of the field each time he plows. As a result the field becomes very slightly higher in the center so that it drains better after a heavy rain.

In 1940 an ingenious young farmer named Frank Andrew devised a way for the tractor to plow his field spirally without a driver. The tractor was guided automatically by means of a stainless steel piano wire attached to the tractor guide bar. The tension was adjustable by means of a spring on the tractor. The other end of the wire was attached to a drum firmly positioned in the center of the field. Each time the tractor went once around the field, one turn of the wire was unwound from the drum, letting the tractor out to a larger curving path. The circumference of the drum was chosen to be just the same as the width of the implement used, whether plow, disc or cultivator, so that the tractor path was shifted by the right amount. For double harrowing, he could start the tractor at the center of the field with the wire just the right length so that it would complete its unwinding when the tractor got to the outside of the field. As the tractor continued on around its path, the wire would start to wind up again and the tractor would spiral in to the center of the field.

The operation may be done at night or when the farmer is away if he puts just enough gasoline in the tank of the tractor to last for the duration of the job. On large level farms, where manpower is at a premium, you may still find Mr. Andrew's ingenious device in use. Mr. Andrew himself is now Professor Andrew of the Department of Agricultural Engineering of the College of Agriculture of the University of Illinois.

The circular fields are increasing in number across the country. As you view them from the air, look for a thin radial line extending from the center to the rim. It is a water-carrying pipe which rides on big wheels and rotates around the center point, sprinkling the field.

Rows of shrubs and small trees separate the fields in the area shown in Plate XXXVIII. Wherever the last ice sheet scattered across the land its heterogeneous load of stones and boulders, you will find such hedgerows between the fields.

Plate XL. Spiral plowing in southern Illinois

The farmer had to remove the boulders from his fields in order to till them and plant his crops. He piled the rocks along the boundary of the field and there the shrubs and trees have grown up among them. Such stone walls with their shrubs and trees trap the drifting snow in the wintertime so that, as the snow melts away from the open areas, the drifts along the stone walls remain as white borders to the brown fields. Flying over such areas as Oklahoma and Arkansas, which were never covered by the continental ice sheet with its load of boulders, you will see hundreds of square miles of farmland spread out below you in an orderly, checkered pattern, field adjoining field with no hedgerows between them because no glacially deposited boulders had to be disposed of by the farmer.

The orderliness of arrangement of the plains from the Appalachian Mountains to the Rockies is evident not only in the patchwork of cultivated fields, but also in the straight roads that meet at right angles. These have an interesting history that dates from 1785 when the Northwest Territory was surveyed and a great rectangular network of reference lines was established. Working from certain arbitrarily chosen "Principal Meridians" and certain "Base Lines" perpendicular to these meridians, the surveyors undertook to lay out the boundaries of a great number of six-mile-square townships. The boundaries were of course parallel to the meridians in the north-south direction and to the Base Lines in the east-west direction. It is these surveyed boundaries that govern the pattern of the roads throughout the north-central United States.

One difficulty encountered by the surveyors arose because of the inescapable fact that meridians of longitude converge northward in the northern hemisphere, meeting at the North Pole. The width of each township at the Base Line was six miles, but about four rows of townships north of the Base Line, the width had become appreciably less than this, as shown in Figure 44. Something had to be done if the townships were to be made approximately six miles square. What the surveyors did was to start over again at a second line, parallel to the Base Line, but about four townships north of it, measuring out the six-mile intervals for the townships along this line as they had along the Base Line. This second line was appropriately called a "correction line." It resulted in the offset of the boundaries shown in Figure 44. When roads were built, they followed the township boundary lines, jogging where they jogged at the correction lines. Although many of these have been smoothed into gentle curves for today's traffic, you can still see from the air some of the correction-line jogs in the roads between Pittsburgh and Denver.

Just west of Chicago, where most of the small towns have

Figure 44. The six-mile-square townships and a correction line.

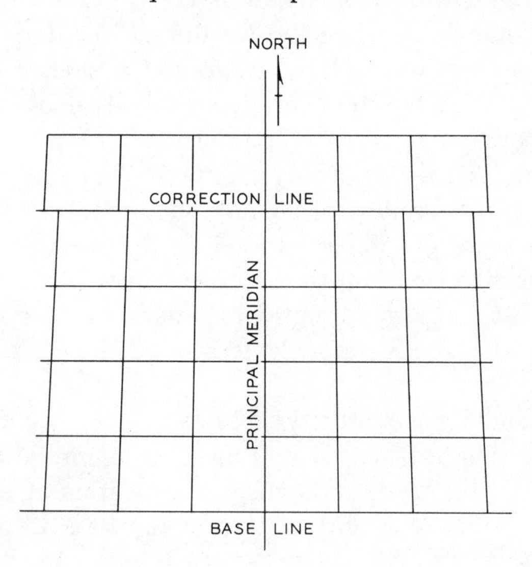

their rectangular street systems neatly aligned with the north-south, east-west grid, there is one with an orderly rectangular street system cocked at an angle of fifteen or twenty degrees to the grid. Peering through the haze to seek an explanation for this irregularity, I soon found it. The railroad tracks had determined the orientation of the town's streets. The inhabitants, used to rectangular street systems, had built such a system parallel and perpendicular to the railroad tracks around which the town developed, even though this meant making many awkward intersections with the adjoining north-south and east-west roads.

In hilly country, roads, like railroads, take advantage of the work of grading that the rivers have performed. Pick a road below you and see whether you can tell why it goes where it goes. Some roads curve to get to a desired place; some roads curve to avoid something. Present-day superhighways, built with the aid of powerful machines which channel through the hills and fill up the hollows, have no sharp curves except their cloverleafed interchanges. You may be able to tell by the radius of the cloverleaf whether the speed limit on the highway below you is 50 or 65 miles per hour.

In Plate XLI we see a broad cleared strip through the woodland. Is it construction for a superhighway or is it a power line strip? By comparison with the two-lane road in the foreground, we can see that it is about four lanes wide. It progresses in a straight line across the land to the middle distance where it turns sharply and then progresses in a straight line again. This marks it as a power transmission line. (If you look closely you may be able to see the towers with their short dark shadows, but these are often invisible from the air.) Highways are built with smooth open curves for long range visibility and minimum skidding when the roads are wet.

Why doesn't the power line curve smoothly? Where the power line runs straight, the horizontal components of the pull of the wires on the opposite sides of any tower exert equal and opposite forces on the tower and do not tend to tip the tower over. Where the direction of the power line changes, there is a resultant force toward the inside of the bend that has to be balanced by a force pulling the tower toward the outside of the bend. This force is generally applied by strong cables attached to firm bases near the tower. This special construction for a tower "on the corner" is more costly than the standard construction of a tower in the straight part of the line. A gradual curve would require several such towers, a single sharp bend requires only one. For this reason power lines, unlike highways, have sharp bends.

The broad clearing around the transmission line is necessary because the electric power is being transmitted at high voltage. If you have walked across a carpeted room on a cold dry day and felt the shock of the transfer of electric charge when you came close to a metal doorknob, you know that the charge can jump a short distance. It does so because there is a difference in charge between you and the doorknob. This difference is measured in volts. When the difference is great, as it is when the voltage is high, the charge can leap a long way. Much cheaper installations could be built if the power were carried at lower voltage, but there is a good reason for high-voltage transmission. When electric current is carried by a wire, some energy is lost to the wire in the form of heat. The amount of energy so transferred is proportional to the resistance of the wire multiplied by the square of the current carried. It is obviously desirable to keep the amount

Plate XLI. A cleared strip: highway construction or power transmission line?

of current low, for minimum energy loss to the wire. The power is proportional to the product of the voltage and the current. The same power can be obtained in a 6-volt headlight lamp in a car, operating at 4 amperes, as in a 120-volt house-lighting circuit at a current of 0.2 ampere. For low losses in electric power transmission, therefore, the low-current power is transmitted at high voltage and then it is transformed, at the end of the transmission line, to the higher-current, lower-voltage electrical power suitable for household use.

Long before the days of power transmission, man knew that a straight line was the shortest distance between two points, and the straight paths worn by the feet of prehistoric man have been discovered from the air by archaeologists in places where they could not be discerned from ground level. If a path is traveled again and again it becomes worn below the level of the land on either side of it. The broad groove may not be noticeable to one standing close to it, but from the air, with the late afternoon sun making long shadows, it may show as a straight line of shadow across a field. Since it is a little below the general surface, it will be less dry than the surrounding land and the vegetation will grow differently in it. This may show better in spring when growth is rapid and slight advantages make a greater difference to the young plants.

An old ditch, filled with looser soil; a buried bit of debris or footing for a wall or dwelling; these will cause different plant growth and so will show differently from the air under suitable conditions. Recognition of these differences takes practice, but the fact that structures made by man form extended patterns is the key to discovery from the air. As D. M. Reeves has put it, in an article on "Aerial Photography and Archeology" in *American Antiquity* (Vol. 2, 1936), "A fly walking on a rug would have difficulty recognizing the design, but upon flying above the rug, the pattern would become distinct."

Archaeologists appreciated the potentialities of aerial photography as early as 1880 and used kites and balloons to raise their cameras above their archaeological sites, but the difficulties were great and the results not very satisfactory. It was the intensive use of air photographs for gaining information about enemy territory in World War I that opened the way for archaeological exploration from the air. One of the earliest examples was the discovery of distinct evidence of former cities on photographs taken in Mesopotamia by the Royal Air Force during World War I. Another famous early discovery was made by Colonel Charles A. Lindbergh and Alfred V. Kidder when they spotted from the air Mayan ruins deep in the jungle of Central America where they had not been known to exist.

In *Flights Over the Ancient Cities of Iran,* a splendid book published by the Oriental Institute of the University of Chicago, Erich F. Schmidt describes the discovery of a group of structures on top of an isolated Persian Mountain. "Rock-hewn chambers, together with basins to gather the water of the winter rains, are visible on the inclined summit of Dadan Mountain. No explorer wandering and searching along normal roads on the face of the earth would be able to see the extraordinary changes made by human labor on the top of this precipitous cliff."

The greatest use of the airplane in archaeology has not, however, been in the discovery of new sites, but in the mapping and extension of old sites and the discovery of unknown features in and around these sites. In the United States, for example, aerial photographs have been used in the study of prehistoric canals in the Gila River valley in Arizona by archaeologists from the Smithsonian Institution. A photograph taken of the well-known archaeological site at Newark, Ohio, showed that evidence of prehistoric occupation extended beyond the supposed boundaries of the site into an area that had been graded for use as an airport.

Perhaps the most famous discovery of this sort was made by the British archaeologist Crawford, about the end of World War I. Archaeologists had found parts of the paths along which the great stones had been dragged for the construction of Stonehenge, but the picture was not complete until the missing portions were detected by Crawford from aerial photographs and their location later confirmed by ground excavations.

Such discoveries are made from moderate elevations. As you fly higher, objects on the ground appear smaller, but you can make them appear larger again by using binoculars. How big does an object have to be to be distinguished by an airplane passenger without binoculars from an elevation of 30,000 feet? To get a feeling for the answer to such a question, let us consider Figure 45. We see that an object 100 feet long on the ground, viewed from an elevation of 100 feet, will subtend an angle of 45° in our field of view.

$$\tan A = \frac{100}{100} = 1$$

$$A = 45^\circ$$

In other words, the light rays reaching the eye of the observer from the left end of the object make an angle of 45° with those coming from the right end of the object, when the eye is over the left end of the object as in Figure 45. From an elevation of 500 feet, how much smaller will that object appear, viewed in the same way? We now have a triangle 500 feet high and 100 feet along the base. The tangent of

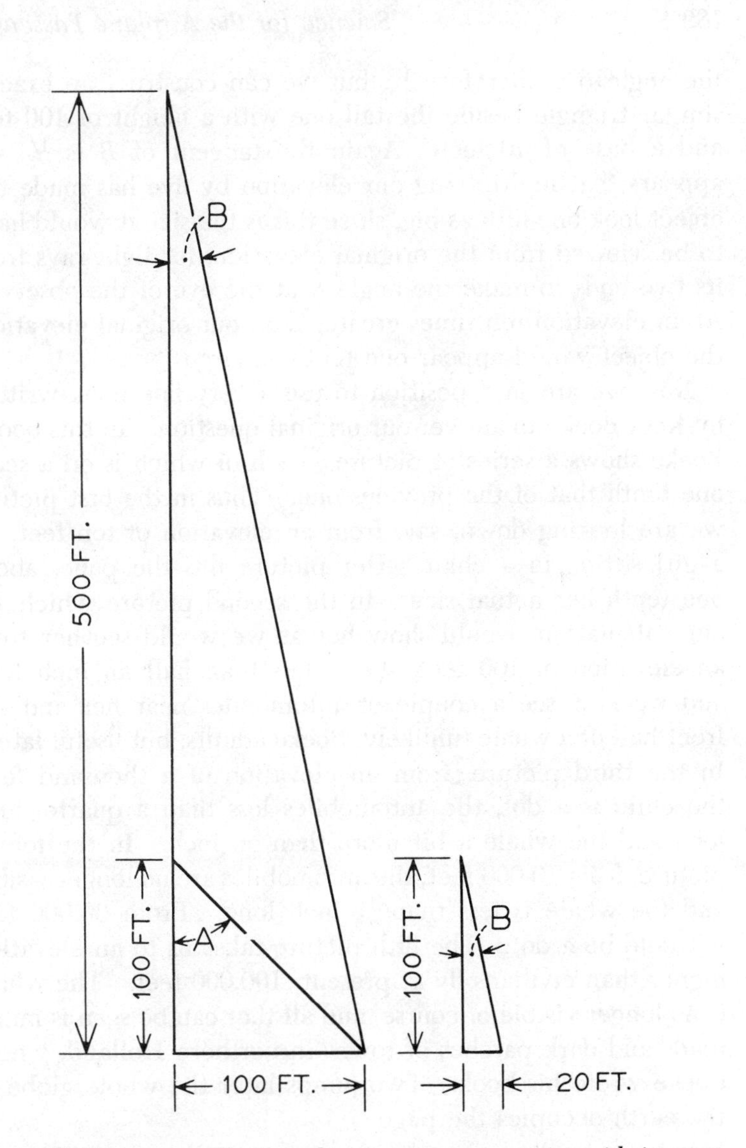

Figure 45. Diagram to illustrate decrease in apparent size with increase in elevation

the angle B is therefore $\frac{1}{5}$, but we can construct an exactly similar triangle beside the tall one with a height of 100 feet and a base of 20 feet. Again the tangent of B is $\frac{1}{5}$. It appears that multiplying our elevation by five has made the object look one fifth as big, since this is the size it would have to be, viewed from the original elevation, for light rays from its two ends to make the angle B at the eye of the observer. At an elevation ten times greater than our original elevation, the object would appear one tenth as large.

Now we are in a position to use a very fine book written by Kees Boeke to answer our original question. In this book* Boeke shows a series of pictures, each of which is on a scale one tenth that of the previous one. Thus in the first picture we are looking down, say, from an elevation of ten feet, on a girl sitting in a chair. Her picture fills the page, about one tenth her actual size. In the second picture which, by our calculation, would show her as we would see her from an elevation of 100 feet, she is less than half an inch long and we now see a couple of automobiles near her and the front half of a whale (unlikely, Boeke admits, but useful later). In the third picture, from an elevation of a thousand feet, the child is a dot, the automobiles less than a quarter-inch long and the whale a bit more than an inch. In the fourth picture, from 10,000 feet, the automobiles are no longer visible and the whale is less than $\frac{1}{8}$ inch long. From 30,000 feet it would be a dot. The fifth picture takes us to an elevation higher than civilians fly at present: 100,000 feet. The whale is no longer visible of course, and all that can be seen is major roads and dark patches of towns in northern Holland, where Boeke wrote his book. Two jumps later the whole globe of the earth occupies the page.

* *Cosmic View, The Universe in Forty Jumps* (New York, The John Day Company, 1957).

What would we conclude about man on the surface of the earth if we looked for signs of his presence from a satellite orbiting the earth? Among those who have turned their attention to this question are Steven Kilston and Carl Sagan of Harvard and Robert Drummond of NASA. Together these three have written a paper entitled, "*A Search for Life on Earth at Kilometer Resolution*"*. By "kilometer resolution" one means that if two objects being photographed were separated by a distance of one kilometer ($5/8$ of a mile), they would appear as separate objects in the photograph. This is a better way to describe the search than to ask what you could detect from a particular elevation, since detection involves the "seeing" and recording equipment as well as the distance to the object to be detected. The abstract of this paper is quoted in its entirety below.

> A search for life on Earth at kilometer resolution, using several thousand photographs obtained by the Tiros and Nimbus meteorological satellites, has been undertaken. No sign of life can be discovered on the vast majority of these photographs. Due principally to the small contrast variations involved and the difficulty in reproducing observing conditions at satellite altitudes, no seasonal variations in the contrast of vegetation could be detected. Of several thousand Nimbus 1 photographs of essentially cloudfree terrains, one feature was found indicative of a technical civilization on Earth — a recently completed interstate highway — and another suggestive feature was discovered, possibly a jet contrail. A striking rectilinear feature was found on the Moroccan coast; however, it appears to be a natural peninsula. An orthogonal grid, discovered in a Tiros 2 photograph, is due to the activities of Canadian loggers, and is a clear sign of life. It appears that several thousand

* *Icarus: International Journal of the Solar System,* Vol. 5 (1966), pp. 79–98.

photographs, each with a resolution of a few tenths of a kilometer, are required before any sign of intelligent life can be found with reasonable reliability. An equivalent Mariner 4 system—taking 22 photographs of the Earth with a resolution of several kilometers—would not detect any sign of life on Earth, intelligent or otherwise.

In the paper one finds that the interstate highway mentioned in the abstract was identified as such only after comparison with a road map of the area and that the feature which might have been a contrail might also have been a natural jetstream cloud. This leaves only the Canadian loggers.

The photographs from the Gemini flights, taken from a little more than 100 miles above the surface of the earth, instead of from the 300- to 600-mile elevation of the Tiros and Nimbus shots, are more revealing of evidence of man on earth. A fine collection of them, in color, has been published in the *National Geographic* (Vol. 130, No. 5, November, 1966, pp. 645–670), accompanying a most interesting article, "The Earth from Orbit", by Dr. Paul D. Lowman, Jr., of NASA's Goddard Space Flight Center. Dr. Lowman describes evidence to show that, under very favorable circumstances, buildings, trains, boat wakes, and perhaps even a large truck were observed by the astronauts. The Egyptian pyramids, however, were missed altogether.

In one of the Gemini 4 photographs accompanying Dr. Lowman's article, the rectangular patterns of farmland in west Texas show clearly and the road between Odessa and Midland shines white in the sunlight.

13. *Flying at Night*

As the daylight fades with the coming of evening, color is drained from the landscape spread out below you. The yellow fields of grain, the green grass, the red barns are still visible in the twilight, but they are no longer so clearly yellow and green and red. We are so used to this phenomenon that we hardly realize that it needs explanation, yet one might reasonably suppose that we should see the colors just as distinctly, if less brightly, as the light grows dim.

The graying of the colors is a physiological effect due to the difference in the way the eye functions in strong light and in weak light. The eye has cone-shaped light-receiving organs, mostly in the center of the retina, and rod-shaped light-receiving organs, mostly around the rim of the retina. The cones are color-distinguishing; they are responsive to light of ordinary intensities, but not to weak light. The rods do not distinguish one color from another, but they are sensitive to light far weaker than that detectable by the cones. As the daylight fades the rods take over where the cones leave off and we buy light sensitivity at the expense of color detection.

In our perception of color we may also have to take into account another physiological effect. If a blue source of light and a red source of light appear equally bright to the eye, and if the intensity of each is diminished in the same ratio (say, cut to one half), then they will no longer appear equally bright. The blue light will appear brighter. This is known as the Purkinje effect, after its discoverer, J. E. Purkinje, who, in the early part of the nineteenth century, also made studies of the nature of vertigo, discovered the germinal vesicle in the chick egg and first pointed out the permanent nature of fingerprints!

With darkness, the lights come on, lights bright enough for our cones to detect their color, lights on bridges and highways, in stores and houses, on signs and automobiles, in airports and on planes. Easily the oldest features of the plane are its red and green lights; red on the left wing and green on the right. Recognizing established custom, the British Admiralty finally wrote into the Rules of the Road in 1848 that ships must carry a red light on the port side and a green light to starboard. Since 1864 such lights have been required on all American ships. The law decrees that they shall be "so fixed as to show the light from right ahead to 22½ degrees (2 points) abaft the beam" (22½ degrees back of a direction 90° to the long axis of the boat). They make it possible to determine in the dark which way a neighboring boat is going. The boat also carries white lights, bow and stern. When only these can be seen, you know the boat is going away from you since the red and green lights cannot be seen from farther back than "2 points abaft the beam".

The lights on boats are steady lights, but some of those on the airplane blink constantly. The boat is seen against a background of black water on which any light means a boat, but the plane is seen against a background of stars, and to avoid collisions it must be distinguishable from them at a great distance, because it moves several miles a minute.

The lights at the airport outline the runways and warn of obstructions. The landing runway is outlined with white or yellow-white lights; the taxiways, along which the airplanes approach or leave the runway, are usually marked with blue lights. Obstructions bear red lights, except that the wind indicator may be lighted in green. The Federal Aviation Agency makes the rules, but there is much variation from one airport to another. For example, some airports have, in addition to their white runway lights, and right beside them, lights which give the pilot information about his approach angle. These lights are so constructed that they show red when viewed along a line of sight making an angle of less than 3° with the ground, but white when viewed at a greater angle. A landing airplane views the near lights at a steeper angle and the far lights at a flatter angle (Fig. 46). If it is viewing the far lights at an angle of less than 3° they will look red. At the same time the near lights will look white if they are being viewed at an angle of greater than 3°. By watching the lights as he approaches, the pilot can judge his angle of approach. If it is correct, he will see the far lights as red and the near lights as white; if too steep, he will see all white; if too low, he will see all red.

Figure 46. Approach-angle lights

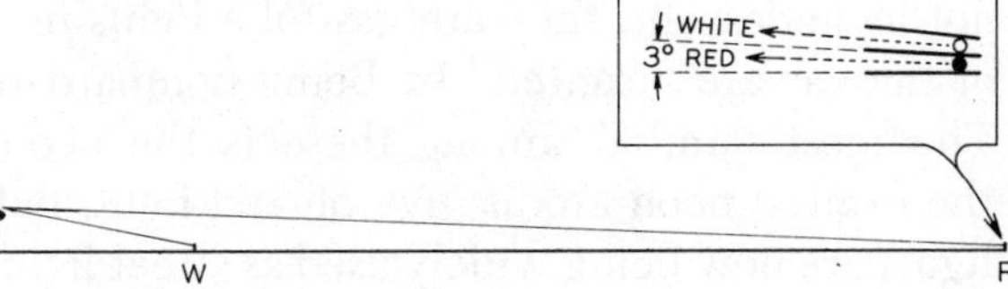

The colored lights in the airport are incandescent lights, like standard household light bulbs, colored by the use of screens or filters or colored lenses.

Light from a very hot incandescent source, such as the hot wire in a common light bulb, has all the various wavelengths to which the human eye is sensitive. The whole rainbow range of colored light beams, taken together, make up white light. When such light is passed through a piece of colored glass, part of it is absorbed by the glass which selectively stops light beams whose wavelengths are in a particular range; the greenish range, for example. Since the light that gets through no longer comprises the whole rainbow range, it is no longer white light. If greenish light has been subtracted from it, it will look reddish.

A spectroscope is an instrument for spreading out the light according to wavelength as it is spread in the rainbow. If light that has been through a red glass filter is examined with a spectroscope, we see that where greenish light should be, there is only darkness because the greenish light did not get through the glass. The heart of a spectroscope, the part that achieves the spreading of the light, may be either a prism or a diffraction grating. Plastic replica diffraction gratings are now widely available at very low cost and are well worth the sliver of space they occupy in your pocket. With such a grating you can analyze light, spread it out into its component beams according to wavelength.

Such an analysis of some of our colored signs and street lights can be of special interest. Many of our city lights are not incandescent; they are gas-tube lights in which the atoms of the gas are "excited" by being bombarded with electrons. The most familiar among these is the neon light in which the excited neon atoms give off red light, but mercury-vapor lights are now being widely used as street lights. In an excited atom, one or more of its electrons has been given extra energy. When the atom returns to its former state this energy is given off as light. The atom can take on energy only in certain specific amounts, different for each different kind of atom. When this specific amount of energy is given off as light, the light is of a specific wavelength, determined by the packet of energy that the atom took on.

When the light from such a source is spread out according to wavelength, it appears only in certain sharply defined narrow regions of the whole rainbow range. Most of the range is dark. If you look at a mercury-vapor street lamp through a plastic replica diffraction grating, you will see, off to the side, a purple street light, separated from a green street light still farther away from the source and a yellow one farther out than that. The excited mercury atoms, returning to their unexcited state can only give off energy as purple light, green light, and yellow light, with no light in between these colors.

Street lights and signs shine directly into our eyes, but house lights cannot be seen directly from the air. Therefore the light that we see from them takes on some of the color of the object that reflects the light to our eyes. In wintertime, if there is no snow on the ground, each house not in a city is marked by a brownish orange spot of light whose color comes from the reddish-brown bare twigs and branches reflecting the house lights. When snow is on the ground each house becomes a starlike thing, with long white fingers of light radiating out from it.

At night the distribution of the population shows more clearly than in the daytime. Houses that might blend into the daytime landscape become pinpoints of light in the night; isolated and lonely looking, or connected, like beads on a string, along a country road. Towns are bright spots with radiating rays, like the spots of color on a tied-and-dyed scarf.

Where the land lies in parallel ridges, as it does in much of Pennsylvania, the lights lie in parallel rows along the valleys. Since these ridges change direction from nearly north-south in southwestern Pennsylvania to nearly east-west in eastern Pennsylvania, you can judge roughly where you are in the state, by consulting the map of your route and observing the angle that the lines of lights below you make with the path of the plane.

In the flat midwestern United States, the rectangular road system lies like a lighted tennis net on the ground, against which the patterns of cities take shape here and there. Some cities have distinctive patterns. Flying over western Pennsylvania one evening, I looked down to discover the brilliantly lighted "golden triangle" of Pittsburgh outlined by the black Allegheny and Monongahela Rivers. There was no doubt about its being Pittsburgh.

The region from Boston to Washington, D.C., has been called "megalopolis," the mammoth metropolis, extending for four hundred miles with no truly inter-city areas. Fly over it at night if you need to be convinced. Here are no bead-strings of lights in the great darkness, no bright spots with rays radiating into the night; just one great interconnecting network of lights with rivers of cars running through it, dammed here and there by toll-booth plazas, beyond which they flow swiftly again.

During the day, air in contact with the sun-warmed earth rises and cools and cumulus clouds form, as discussed in Chapter 6. Early in the evening these clouds disappear because they depend on the rising, cooling air for their nourishment. Since these convection currents are responsible for low-level bumpiness (Fig. 21), night flying at low levels is generally smoother than day flying at these levels.

If you are flying at night in northern latitudes, especially during the months of March, April, September and October, you have a good chance of seeing that magnificent spectacle, the aurora borealis. According to Neuberger*, "An average of about 240 nights with auroral displays a year can be observed along a line passing through the lower half of Hudson Bay, near Point Barrow, Alaska, through the northern portion of Siberia and Norway, through Iceland, and slightly off the southern tip of Greenland. . . . In the northern half of the United States, 5 to 25 nights per year show some northern lights." Those who have seen these lights from the particularly advantageous observation post provided by a high-flying airplane recall the experience with awed delight.

The aurora occurs very high in the earth's atmosphere, above the stratosphere which is above the troposphere in which we live. In the middle latitudes in summer the troposphere may be eleven or more miles deep, but in winter it may be less than half that deep. Planes flying at about 30,000 feet are commonly in the upper troposphere, but in winter they may be above the troposphere, in the stratosphere. The still, clear air of the stratosphere extends to approximately thirty miles above the earth's surface. Above this, the atoms of the very thin air are subjected to the action of cosmic rays and ultraviolet radiation from the sun which strip from some of them their outermost electrons, thus changing them into *ions* with a net positive charge. This region of the atmosphere is therefore called the *ionosphere.*

According to the theory of the aurora most favored at present**, streams of particles, probably electrons, from the sun bombard the ions and atoms of the ionosphere, giving

*Hans Neuberger, *Introduction to Meteorology,* (University Park, Penna., The Pennsylvania State University Press, 1965).

**Neuberger, op. cit.

them enough energy so that they become excited. As each one subsequently returns to its unexcited state it gives off the energy it received, emitting it in the form of light of a particular wavelength, determined by the nature of the ion or atom, just as the atoms of neon gas do in a neon light tube. An analysis of the light emitted tells us about the atoms that are emitting it. The greenish light commonly observed in the aurora comes from excited oxygen atoms.

The electron-stream theory is consistent with the fact that the aurora is most commonly seen near the north and south magnetic poles. (These auroras are called the *aurora borealis* and *aurora australis*, respectively.) According to the theory, the earth's magnetic field deflects the stream of charged particles from the sun toward the magnetic polar regions where they descend along the "lines of magnetic force", swarming through the ionosphere and causing the aurora. Below the ionosphere they are stopped by the more closely spaced molecules of the denser air.

The elevation of the aurora above the surface of the earth is most commonly between 60 and 200 miles. How big is it? This varies from one display to the next, but you may be able to get an order-of-magnitude estimate of the size of any particular aurora you have the good fortune to see by using your bookback goniometer. Suppose you find, by the method shown in Figure 16, that some great green veil in the northern sky subtends an angle, A, of 20°. Then you know that its length, *L*, (Fig. 47), is related to its distance from your eye, *D*, by

$$\text{Tangent A} = \tan 20^\circ = \frac{\text{L}}{\text{D}}$$

You can get the tangent of 20° from Table 2 (page 110), but what is the value of *D*? We can make a rough estimate of it by assuming that the bottom of the aurora is about 60 miles above the surface, the elevation most commonly observed. Then, by measuring the angle between the horizon and the bottom of the aurora (angle *B*, Fig. 47) we can calculate *D* from the relationship

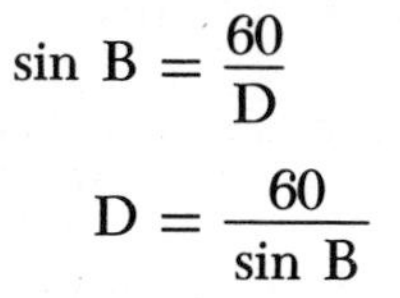

$$\sin B = \frac{60}{D}$$

$$D = \frac{60}{\sin B}$$

Figure 47. Measuring the aurora

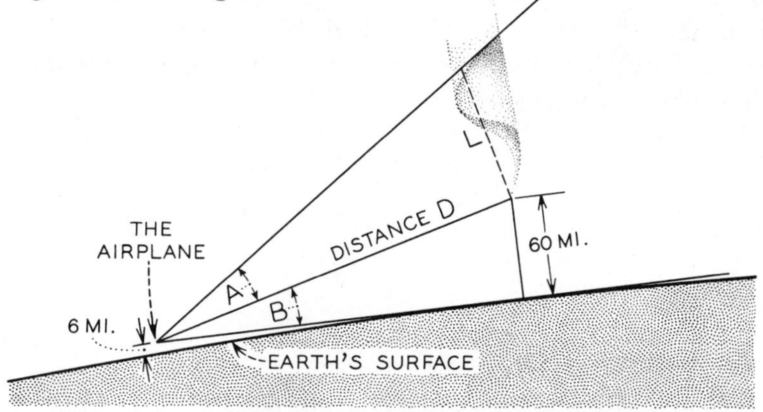

A sine table is given below for your convenience.

Table 4
Sines of some small angles

Angle	*Sine*
5°	.09
10	.17
15	.26
20	.34
25	.42
30	.50

Suppose you find the angle of elevation (*B*, Fig. 48) to be 15°. Then the estimated distance of the aurora from you is

$$D = \frac{60}{.26} = \text{approximately 230 miles}$$

From this you can determine that the length of the aurora is

$$L = D \tan 20° = 230 \times .36 = \text{approximately 83 miles.}$$

Since we used the right triangle involving the angle, *A*, and the side, *D*, in Figure 47, the distance, *L*, we have calculated is in fact the length of the dashed line in the figure. The vertical length of the aurora may be a little greater or less, depending on its shape.

14. Landing

An airplane several miles above the surface of the earth starts its descent many miles before its destination. You may be able to sense when this happens by your inertial frame of reference or by hearing the change in sound of the motors or jets. Or the captain may tell you about it: "We have just passed over Easton, Pennsylvania, and are beginning our descent for Newark Airport." Suppose you have been cruising at 31,000 feet (5.9 miles) when you hear this announcement. Reference to a map that has a scale of miles will show you that Easton is 55 miles from Newark Airport. From this information you can calculate the average angle of descent, *D*; the average angle that the airplane's path makes with the horizontal as it descends.

$$\text{tangent D} = \frac{5.9}{55} = 0.107$$

From Table 2, you can see that *D* is a little greater than 6°.

Do all airplanes descend at the same average angle or do

some drop out of the sky more steeply than others? If you keep track of your location carefully as you near the end of your trip by spotting recognizable features of the land below you and matching them to features on a map, you will know where you are when the plane starts on its long downward path. By doing this on several trips in different types of airplanes and keeping records of the descent angles, you will be able to answer the question from your own observations. The Federal Aviation Agency requires that the angle be 3° for the last 4 miles of descent.

As you descend, you may again sense discomfort in your ears. Babies that have been quiet during the trip protest about this noisily. This time the pressure in the middle ear (M, Fig. 3) is less than the outside pressure and the resulting adjustment of the surrounding tissue is felt by you. Air must move into the middle ear along the Eustachian tube if this uncomfortable sensation is to be relieved, but the pressure of the outside air tends to keep it closed and the discomfort is therefore likely to be longer lasting than it was on the way up.

If you approach the airport from any but the downwind side, the airplane will have to head for a point downwind from the airport and turn into the wind to land. Its lift comes from the motion of the air past the wings (Chapter 1) and this must be kept as large as possible when the plane slows down for a landing. Yet if the plane's velocity relative to the landing strip is too great, it will have difficulty in stopping before the end of the strip. So it must turn to land facing the wind and it must *bank* as it turns.

You are familiar with banked turns on roadways on the ground. If you are driving fast on a wet paved road and come to an unbanked curve, you know you are likely to skid toward the outside of the curve because of the tendency of the car to continue its motion in a straight line. When the car is standing still or moving at constant velocity in a straight line, the only force on it is the gravitational attraction between it and the earth. This force is the car's weight, the force with which it presses down on the pavement. The pavement presses up on the car with an equal and opposite force. If this were not so, there would be a net force one way or the other and the contact between the pavement and the tire would move up or down in response to the net force.

When the wheels are turned as we go around a curve, we are using the friction between the tire and the pavement to exert the necessary force on the car (*f*, Fig. 48) to divert it from its tendency to persevere "in its state of uniform motion in a straight line" (Chapter 1). If, however, the friction is decreased by a lubricating layer of water or ice, then the turning force will not be able to act and the car will continue

Figure 48. Top view of a car going around a curve on a dry, unbanked road

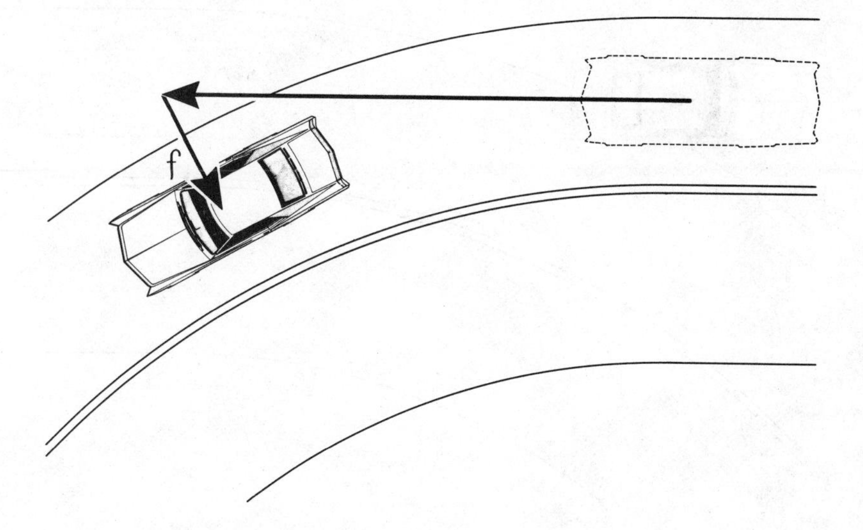

to move in a straight line toward the outside of the curve in accordance with Newton's first law (Fig. 49). Such a skid can be disastrous.

In Figure 50, showing the tire of a car in relation to the pavement, its tendency to move toward the outside of the curve (that is, to persevere in a straight line) is overcome by the force f toward the inside of the curve. (The arrow W in Fig. 50 represents the weight of the car, the force which is opposed by an equal and opposite upward force F from the pavement.) If the force f cannot act (Fig. 51) because glare ice has reduced the road friction to zero, then, as we saw in Fig. 49, the car will continue in a straight line and the tire will skid toward the outside of the turn.

Suppose the road is banked as in Fig. 52. Then the weight of the car no longer acts in a direction perpendicular to the

Figure 49. Top view of a car failing to make the curve on an icy, unbanked road

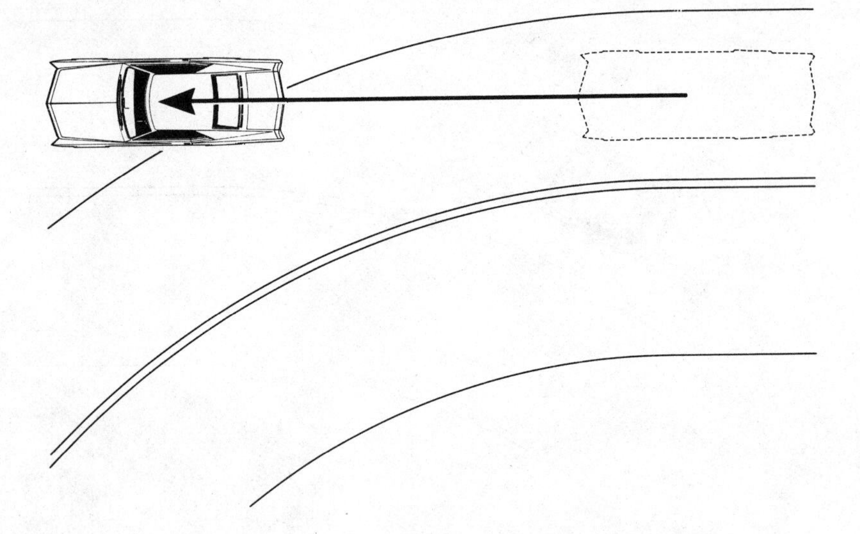

road, but at an angle to it which is equal to the banking angle, *B*, of the road (Fig. 52). This force due to the weight of the car can be analyzed into two components, one perpendicular to the pavement (*p*, Fig. 52) and one along the pavement (*a*, Fig. 52), since these two forces, acting together, would have the same effect as the resultant force *W*. The force *p* is opposed by an equal and opposite force *F*, on the tire by the pavement, as before, and this leaves us with the force *a*, which is just what we need to counteract the tendency of the tire to move toward the outside of the curve as it continues in a straight line (Fig. 49).

The correct banking angle, *B* (Fig. 52), will depend on the speed of the car and the sharpness of the curve. Suppose the road is icy and the banking angle is not steep enough: the car will skid toward the outside of the curve because the force *a* is

Figure 50. Left, tire going around a curve to the left on a dry, unbanked road

Figure 51. Right, tire failing to make the curve to the left on an icy, unbanked road

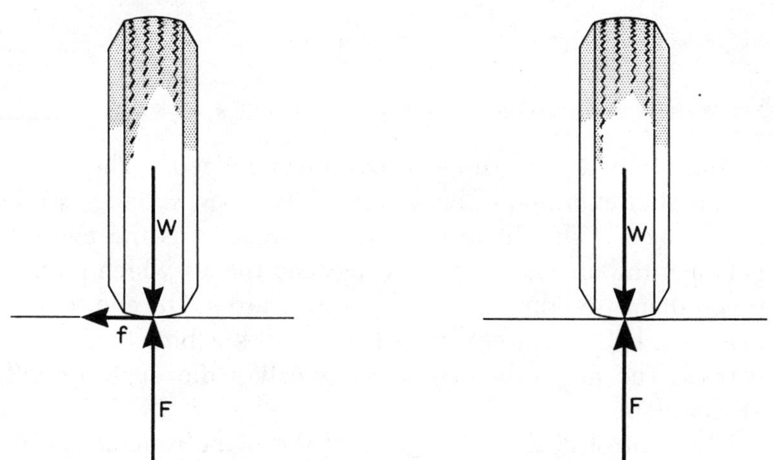

not big enough to deflect it adequately from its straight-line path. If the banking angle is too steep, the car will slip downhill toward the inside of the curve because the force *a* (Fig. 52) is bigger than that needed to deflect it from its straight-line path. When the angle is exactly appropriate to the speed of the car and the sharpness of the curve, the car can make the curve on glare ice, with no tendency to slip either toward the outside or the inside of the curve.

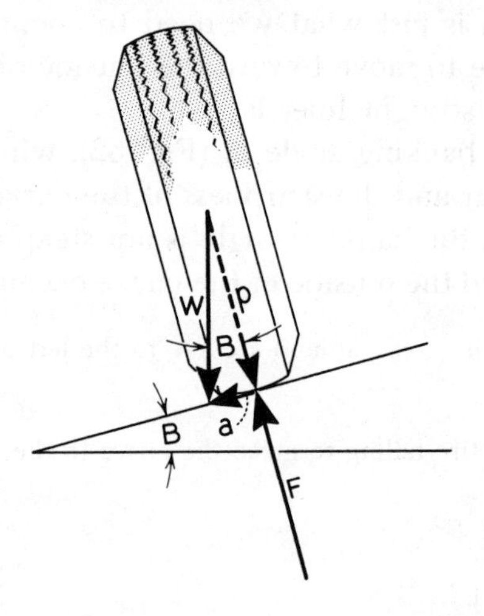

Figure 52. Tire going around a curve to the left on a banked road

The air is a very slippery pavement indeed. The plane would skid enormously on a turn if its "road surface" were not banked. The "road surface" in question is the contact between the underside of the wings and the air which pushes up on them. So the plane banks, tilting around its end-to-end axis as a baby's cradle tilts. Just so does a bird bank as it turns. The angle of such a tilt is called the angle of *roll* of the plane.

The control of the movements of the plane is achieved by

means of hinged flaps. The one attached to the vertical fin of the tail is called the *rudder*; those on the trailing edges of the horizontal tail surfaces are called *elevators;* and on the trailing edge of each wing is a hinged flap called an *aileron.* When the rudder is straight out behind, the air flows smoothly past it, but when the free rear edge of the rudder is swung to the right, the air moving past the right side of the plane pushes against it and so pushes the tail of the plane to the left; therefore the plane turns right. This motion (a rotation around a vertical axis) is known as *yaw.* The elevators, together with the horizontal part of the tail, form an air foil like that in Figure 1. Lowering the free rear edge of the elevators increases the curvature of the air foil and lifts the tail, causing the nose of the plane to be lower than the tail. This is the motion we have already identified as pitch. The controls for the ailerons are so linked that when the right-wing aileron is raised, the left wing aileron is automatically lowered. The lowering of the left aileron increases the effective curvature of the wing-aileron assembly and causes lifting of the left wing, while the right wing goes down because of the opposite action and the plane banks to the right. In this way the ailerons control the roll of the plane.

Can you measure the angle of roll of the plane, using the bookback goniometer as a clinometer as you did for the angle of *pitch* in Chapter 9 (Fig. 34)? Use the back of the seat in front of you as a reference line in the plane which would be horizontal when the angle of roll was zero. Try it as the plane goes into its banked turn just prior to landing. You may be surprised at the result of your measurement. If you are, remember Newton's first law and the discussion of Fig. 52.

As the plane comes close to the runway it must cut its speed drastically, but it still needs the lift that it gets from its speed relative to the air. A strong headwind can help at this stage, but the crew can also help a little. If the curvature of the upper surface of both wings could be increased, the lifting force would be increased. This is done by extending curved flaps out from the rear edge of the wing. These also help to slow down the plane but their primary purpose is to increase the lift at low velocity.

After landing, the speed of the plane must be decreased drastically. In the early days of flying, brakes on the wheels had to do the whole job, but in today's planes, both propeller and jet, the direction of the engines' thrust can be reversed. In the propeller plane this is done by changing the angle of the propeller blades so that they blow air forward instead of backward. In a jet, deflector buckets catch the jet exhaust and deflect it forward. The plane experiences an equal and opposite force. As Newton put it, in his third law, "To every action there is always opposed an equal reaction: or, the mutual actions of two bodies upon each other are always equal, and directed in contrary parts."

In the old way, when we had to depend on wheel brakes, if the runway was slippery there might not be enough friction between the tires and the runway to apply enough force to prevent the plane from "persevering in its state of uniform motion" in accordance with Newton's first law. In the new way we are not depending on the interaction between the wheels and the runway, but on the interaction between the air moved by the motors and the air surrounding the plane, a safer dependence.

Your airplane may turn sharply from the landing runway onto a taxi lane. This turn is not banked at all since most of it is on the runway itself which must be flat. What happens in this case when you hold the clinometer in the left-to-right position, with one edge against the horizontal edge of the chairback in front of you? Compare this case with the measurement made during the banking turn in the light of your knowledge of the orientation of the forces in each case.

The buildings at airports are as varied as the cities they serve. Some are made of logs, some of concrete, some of glass and chromium-plated steel. But somewhere in every airport will be the familiar shape of the heart of the control tower, the top-heavy "control cab" with the sloping green glass windows whose roof bristles with assorted antennae (Plate XLII). This doesn't happen by chance. Technical Standard Order N136 of the Federal Aviation Administration specifies rather precisely how the control cab shall be built. For example, it requires that "to minimize reflections in the control cab, the glass shall be sloped outward from sill to ceiling at an angle of 15 degrees from the vertical." You can check this angle with the bookback goniometer, using a weighted string or chain as in Figure 34 to show you the vertical direction. If the window panes were vertical, beams of the late

Plate XLII. The control cab

afternoon sunlight that entered the control cab might be reflected from the inner surface of the panes into the eyes of those watching the airplanes; with the panes sloping outward, such beams are reflected upward to the ceiling. With vertical windows sunlight might also be reflected into the eyes of landing pilots; the outer surfaces of the sloping panes reflect light downward to the ground.

TSON136 further states that "double glass with a hermetically-sealed air space shall be installed in the tower cab because of its superior insulating value to minimize fogging or frosting of the glass." The air in the control cab must not reach the dew point. The air between the outer pane and the inner pane is a poorer conductor of heat than the glass is, so the inner glass does not lose its heat as rapidly to the outside and no water droplets condense on it to hide the landing planes from view.

TSON136 goes on, "a tinted heat-absorbing plate glass with a light transmittance value of not less than 70% shall be used as the outer pane only, the inner pane being clear."

On top of the control tower shown in Plate XLII are three different kinds of antennae, each with a different function. The two vertical whip-like antennae at the left are for communicating with private aircraft and aircraft maintenance crews. The oscillating electric current moving up and down in these vertical rods will send out radio waves whose oscillation is only up and down, polarized radio waves which must be received on vertically oriented antennae. The vertical rod that extends out of the picture, topping everything, is a lightning rod. The two large irregular antennae, which have been nicknamed *swastikas* by the airmen because of their shape, transmit radio waves which can be received on either vertical or horizontal antennae. The four antennae on the four corners of the control cab, the bristling antennae that look like witches' brooms, are for communication with military aircraft. Elsewhere on the field there is a horizontal antenna for transmitting navigation information which always goes out on horizontally oscillating radio waves.

We have come to the end of the trip. We have observed many things, but there are many more still to be observed. We have made a few measurements, but you will think of others that can be made. We have discovered, as airplane passengers, much that we could not have observed from the ground. Watch out for this game of observing the world around you. It is habit-forming.

Suggested Reading

Suggested Reading

(Paperbacks are marked with an asterisk)

THE AIRPLANE AND ITS OPERATION:

Federal Aviation Agency, Flight Standards Service, *Private Pilot's Handbook of Aeronautical Knowledge.* Washington, U.S. Gov't. Printing Office, 1965.

Murchie, Guy, *Song of the Sky.* Boston, Houghton Mifflin Company, 1963.

ARCHAEOLOGY:

St. Joseph, J. K. S., ed., *The Uses of Air Photography, Nature and Man in a New Perspective.* London, John Baker Publishers, Ltd., 1966.

Schmidt, Eric F., *Flights over the Ancient Cities of Iran.* Chicago, University of Chicago Press, 1940.

CLOUDS AND METEOROLOGY:

Federal Aviation Agency, Flight Standards Service, *Private Pilot's Handbook of Aeronautical Knowledge.* Washington, U.S. Gov't. Printing Office, 1965.

* Lehr, Paul E., Burnett, R. Will, and Zim, Herbert S., *Weather.* New York, Golden Press Inc., 1963.

* Mason, B. J., *Clouds, Rain and Rainmaking.* Cambridge, England, Cambridge University Press, 1962.

Murchie, Guy, *Song of the Sky.* Boston, Houghton Mifflin Company, 1963.

Neuberger, Hans, *Introduction to Physical Meteorology* (for more advanced study). University Park, Penna., The Pennsylvania State University Press, 1965.

Riehl, Herbert, *Tropical Meteorology.* New York, McGraw-Hill Book Company, 1954.

GEOLOGY AND GEOGRAPHY:

Bascom, Willard, "Ocean Waves", *Scientific American,* August, 1959, p. 74.

Leopold, L. B. and Langbein, W. B., "River Meanders", *Scientific American,* June, 1966, pp. 60–70.

Lobeck, A. K., *Airways of America.* New York, The Geographical Press, Columbia University Press, 1933.

Ray, Richard G., *Aerial Photographs in Geological Interpretation and Mapping: U.S.G.S. Professional Paper 373.* Washington, U.S. Gov't. Printing Office, 1960.

St. Joseph, J. K. S., ed., *The Uses of Air Photography, Nature and Man in a New Perspective.* London, John Baker Publishers, Ltd., 1966.

Shelton, John S., *Geology Illustrated.* San Francisco, W. H. Freeman and Company, 1966.

Strahler, A. N., *Introduction to Physical Geography.* New York, N.Y., John Wiley & Sons, Inc., 1965.

LIGHT:

* Humphreys, W. J., *Physics of the Air.* New York, Dover Publications, Inc., 1964.

* Minnaert, M., *The Nature of Light and Colour in the Open Air.* New York, Dover Publications, Inc., 1954.

SCALE OF OBJECTS FROM VARIOUS ELEVATIONS:

Boeke, Kees, *Cosmic View, the Universe in 40 Jumps.* New York, The John Day Company, Inc., 1957.

Kilston, S., Sagan, C. and Drummond, R., "A Search for Life on Earth at Kilometer Resolution", *Icarus,* International Journal of the Solar System, Vol. 5, 1966, p. 79.

Lowman, Paul D., Jr., "The Earth from Orbit", *National Geographic,* Vol. 130, No. 5, November, 1966, p. 645.

STORMS:

* Battan, Louis J., *The Nature of Violent Storms.* Garden City, N.Y., Anchor Books, Doubleday & Company, Inc., 1961.

* Battan, Louis J., *The Thunderstorm.* New York, Signet Key Books, 1964.

Index

Index

A CATALOGUE OF SELECTED DOVER BOOKS IN ALL FIELDS OF INTEREST

A CATALOGUE OF SELECTED DOVER BOOKS IN ALL FIELDS OF INTEREST

DRAWINGS OF WILLIAM BLAKE, William Blake. 92 plates from Book of Job, *Divine Comedy, Paradise Lost,* visionary heads, mythological figures, Laocoon, etc. Selection, introduction, commentary by Sir Geoffrey Keynes. 178pp. 8⅛ x 11. 22303-5 Pa. $4.00

ENGRAVINGS OF HOGARTH, William Hogarth. 101 of Hogarth's greatest works: *Rake's Progress, Harlot's Progress, Illustrations for Hudibras, Before and After, Beer Street and Gin Lane,* many more. Full commentary. 256pp. 11 x 13¾. 22479-1 Pa. $12.95

DAUMIER: 120 GREAT LITHOGRAPHS, Honore Daumier. Wide-ranging collection of lithographs by the greatest caricaturist of the 19th century. Concentrates on eternally popular series on lawyers, on married life, on liberated women, etc. Selection, introduction, and notes on plates by Charles F. Ramus. Total of 158pp. 9⅜ x 12¼. 23512-2 Pa. $5.50

DRAWINGS OF MUCHA, Alphonse Maria Mucha. Work reveals draftsman of highest caliber: studies for famous posters and paintings, renderings for book illustrations and ads, etc. 70 works, 9 in color; including 6 items not drawings. Introduction. List of illustrations. 72pp. 9⅜ x 12¼. (Available in U.S. only) 23672-2 Pa. $4.00

GIOVANNI BATTISTA PIRANESI: DRAWINGS IN THE PIERPONT MORGAN LIBRARY, Giovanni Battista Piranesi. For first time ever all of Morgan Library's collection, world's largest. 167 illustrations of rare Piranesi drawings—archeological, architectural, decorative and visionary. Essay, detailed list of drawings, chronology, captions. Edited by Felice Stampfle. 144pp. 9⅜ x 12¼. 23714-1 Pa. $7.50

NEW YORK ETCHINGS (1905-1949), John Sloan. All of important American artist's N.Y. life etchings. 67 works include some of his best art; also lively historical record—Greenwich Village, tenement scenes. Edited by Sloan's widow. Introduction and captions. 79pp. 8⅜ x 11¼. 23651-X Pa. $4.00

CHINESE PAINTING AND CALLIGRAPHY: A PICTORIAL SURVEY, Wan-go Weng. 69 fine examples from John M. Crawford's matchless private collection: landscapes, birds, flowers, human figures, etc., plus calligraphy. Every basic form included: hanging scrolls, handscrolls, album leaves, fans, etc. 109 illustrations. Introduction. Captions. 192pp. 8⅞ x 11¾. 23707-9 Pa. $7.95

DRAWINGS OF REMBRANDT, edited by Seymour Slive. Updated Lippmann, Hofstede de Groot edition, with definitive scholarly apparatus. All portraits, biblical sketches, landscapes, nudes, Oriental figures, classical studies, together with selection of work by followers. 550 illustrations. Total of 630pp. 9⅛ x 12¼. 21485-0, 21486-9 Pa., Two-vol. set $15.00

THE DISASTERS OF WAR, Francisco Goya. 83 etchings record horrors of Napoleonic wars in Spain and war in general. Reprint of 1st edition, plus 3 additional plates. Introduction by Philip Hofer. 97pp. 9⅜ x 8¼. 21872-4 Pa. $3.75

THE EARLY WORK OF AUBREY BEARDSLEY, Aubrey Beardsley. 157 plates, 2 in color: *Manon Lescaut, Madame Bovary, Morte Darthur, Salome,* other. Introduction by H. Marillier. 182pp. 8⅛ x 11. 21816-3 Pa. $4.50

THE LATER WORK OF AUBREY BEARDSLEY, Aubrey Beardsley. Exotic masterpieces of full maturity: *Venus and Tannhauser, Lysistrata, Rape of the Lock, Volpone,* Savoy material, etc. 174 plates, 2 in color. 186pp. 8⅛ x 11. 21817-1 Pa. $4.50

THOMAS NAST'S CHRISTMAS DRAWINGS, Thomas Nast. Almost all Christmas drawings by creator of image of Santa Claus as we know it, and one of America's foremost illustrators and political cartoonists. 66 illustrations. 3 illustrations in color on covers. 96pp. 8⅜ x 11¼. 23660-9 Pa. $3.50

THE DORÉ ILLUSTRATIONS FOR DANTE'S DIVINE COMEDY, Gustave Doré. All 135 plates from Inferno, Purgatory, Paradise; fantastic tortures, infernal landscapes, celestial wonders. Each plate with appropriate (translated) verses. 141pp. 9 x 12. 23231-X Pa. $4.50

DORÉ'S ILLUSTRATIONS FOR RABELAIS, Gustave Doré. 252 striking illustrations of *Gargantua and Pantagruel* books by foremost 19th-century illustrator. Including 60 plates, 192 delightful smaller illustrations. 153pp. 9 x 12. 23656-0 Pa. $5.00

LONDON: A PILGRIMAGE, Gustave Doré, Blanchard Jerrold. Squalor, riches, misery, beauty of mid-Victorian metropolis; 55 wonderful plates, 125 other illustrations, full social, cultural text by Jerrold. 191pp. of text. 9⅜ x 12¼. 22306-X Pa. $7.00

THE RIME OF THE ANCIENT MARINER, Gustave Doré, S. T. Coleridge. Dore's finest work, 34 plates capture moods, subtleties of poem. Full text. Introduction by Millicent Rose. 77pp. 9¼ x 12. 22305-1 Pa. $3.50

THE DORE BIBLE ILLUSTRATIONS, Gustave Doré. All wonderful, detailed plates: Adam and Eve, Flood, Babylon, Life of Jesus, etc. Brief King James text with each plate. Introduction by Millicent Rose. 241 plates. 241pp. 9 x 12. 23004-X Pa. $6.00

THE COMPLETE ENGRAVINGS, ETCHINGS AND DRYPOINTS OF ALBRECHT DURER. "Knight, Death and Devil"; "Melencolia," and more—all Dürer's known works in all three media, including 6 works formerly attributed to him. 120 plates. 235pp. 8⅜ x 11¼. 22851-7 Pa. $6.50

MAXIMILIAN'S TRIUMPHAL ARCH, Albrecht Dürer and others. Incredible monument of woodcut art: 8 foot high elaborate arch—heraldic figures, humans, battle scenes, fantastic elements—that you can assemble yourself. Printed on one side, layout for assembly. 143pp. 11 x 16. 21451-6 Pa. $5.00

THE COMPLETE WOODCUTS OF ALBRECHT DURER, edited by Dr. W. Kurth. 346 in all: "Old Testament," "St. Jerome," "Passion," "Life of Virgin," Apocalypse," many others. Introduction by Campbell Dodgson. 285pp. 8½ x 12¼. 21097-9 Pa. **$7.50**

DRAWINGS OF ALBRECHT DURER, edited by Heinrich Wolfflin. 81 plates show development from youth to full style. Many favorites; many new. Introduction by Alfred Werner. 96pp. 8⅛ x 11. 22352-3 Pa. $5.00

THE HUMAN FIGURE, Albrecht Dürer. Experiments in various techniques—stereometric, progressive proportional, and others. Also life studies that rank among finest ever done. Complete reprinting of *Dresden Sketchbook.* 170 plates. 355pp. 8⅜ x 11¼. 21042-1 Pa. $7.95

OF THE JUST SHAPING OF LETTERS, Albrecht Dürer. Renaissance artist explains design of Roman majuscules by geometry, also Gothic lower and capitals. Grolier Club edition. 43pp. 7⅞ x 10¾ 21306-4 Pa. $3.00

TEN BOOKS ON ARCHITECTURE, Vitruvius. The most important book ever written on architecture. Early Roman aesthetics, technology, classical orders, site selection, all other aspects. Stands behind everything since. Morgan translation. 331pp. 5⅜ x 8½. 20645-9 Pa. $4.50

THE FOUR BOOKS OF ARCHITECTURE, Andrea Palladio. 16th-century classic responsible for Palladian movement and style. Covers classical architectural remains, Renaissance revivals, classical orders, etc. 1738 Ware English edition. Introduction by A. Placzek. 216 plates. 110pp. of text. 9½ x 12¾. 21308-0 Pa. $10.00

HORIZONS, Norman Bel Geddes. Great industrialist stage designer, "father of streamlining," on application of aesthetics to transportation, amusement, architecture, etc. 1932 prophetic account; function, theory, specific projects. 222 illustrations. 312pp. 7⅞ x 10¾. 23514-9 Pa. $6.95

FRANK LLOYD WRIGHT'S FALLINGWATER, Donald Hoffmann. Full, illustrated story of conception and building of Wright's masterwork at Bear Run, Pa. 100 photographs of site, construction, and details of completed structure. 112pp. 9¼ x 10. 23671-4 Pa. **$5.50**

THE ELEMENTS OF DRAWING, John Ruskin. Timeless classic by great Viltorian; starts with basic ideas, works through more difficult. Many practical exercises. 48 illustrations. Introduction by Lawrence Campbell. 228pp. 5⅜ x 8½. 22730-8 Pa. $3.75

GIST OF ART, John Sloan. Greatest modern American teacher, Art Students League, offers innumerable hints, instructions, guided comments to help you in painting. Not a formal course. 46 illustrations. Introduction by Helen Sloan. 200pp. 5⅜ x 8½. 23435-5 Pa. **$4.00**

CATALOGUE OF DOVER BOOKS

THE ANATOMY OF THE HORSE, George Stubbs. Often considered the great masterpiece of animal anatomy. Full reproduction of 1766 edition, plus prospectus; original text and modernized text. 36 plates. Introduction by Eleanor Garvey. 121pp. 11 x 14¾. 23402-9 Pa. $6.00

BRIDGMAN'S LIFE DRAWING, George B. Bridgman. More than 500 illustrative drawings and text teach you to abstract the body into its major masses, use light and shade, proportion; as well as specific areas of anatomy, of which Bridgman is master. 192pp. 6½ x 9¼. (Available in U.S. only) 22710-3 Pa. $3.50

ART NOUVEAU DESIGNS IN COLOR, Alphonse Mucha, Maurice Verneuil, Georges Auriol. Full-color reproduction of *Combinaisons ornementales* (c. 1900) by Art Nouveau masters. Floral, animal, geometric, interlacings, swashes—borders, frames, spots—all incredibly beautiful. 60 plates, hundreds of designs. 9⅜ x 8-1/16. 22885-1 Pa. $4.00

FULL-COLOR FLORAL DESIGNS IN THE ART NOUVEAU STYLE, E. A. Seguy. 166 motifs, on 40 plates, from *Les fleurs et leurs applications decoratives* (1902): borders, circular designs, repeats, allovers, "spots." All in authentic Art Nouveau colors. 48pp. 9⅜ x 12¼. 23439-8 Pa. $5.00

A DIDEROT PICTORIAL ENCYCLOPEDIA OF TRADES AND INDUSTRY, edited by Charles C. Gillispie. 485 most interesting plates from the great French Encyclopedia of the 18th century show hundreds of working figures, artifacts, process, land and cityscapes; glassmaking, papermaking, metal extraction, construction, weaving, making furniture, clothing, wigs, dozens of other activities. Plates fully explained. 920pp. 9 x 12. 22284-5, 22285-3 Clothbd., Two-vol. set $40.00

HANDBOOK OF EARLY ADVERTISING ART, Clarence P. Hornung. Largest collection of copyright-free early and antique advertising art ever compiled. Over 6,000 illustrations, from Franklin's time to the 1890's for special effects, novelty. Valuable source, almost inexhaustible.
Pictorial Volume. Agriculture, the zodiac, animals, autos, birds, Christmas, fire engines, flowers, trees, musical instruments, ships, games and sports, much more. Arranged by subject matter and use. 237 plates. 288pp. 9 x 12. 20122-8 Clothbd. $14.50

Typographical Volume. Roman and Gothic faces ranging from 10 point to 300 point, "Barnum," German and Old English faces, script, logotypes, scrolls and flourishes, 1115 ornamental initials, 67 complete alphabets, more. 310 plates. 320pp. 9 x 12. 20123-6 Clothbd. $15.00

CALLIGRAPHY (CALLIGRAPHIA LATINA), J. G. Schwandner. High point of 18th-century ornamental calligraphy. Very ornate initials, scrolls, borders, cherubs, birds, lettered examples. 172pp. 9 x 13. 20475-8 Pa. $7.00

ART FORMS IN NATURE, Ernst Haeckel. Multitude of strangely beautiful natural forms: Radiolaria, Foraminifera, jellyfishes, fungi, turtles, bats, etc. All 100 plates of the 19th-century evolutionist's *Kunstformen der Natur* (1904). 100pp. 9⅜ x 12¼. 22987-4 Pa. $5.00

CHILDREN: A PICTORIAL ARCHIVE FROM NINETEENTH-CENTURY SOURCES, edited by Carol Belanger Grafton. 242 rare, copyright-free wood engravings for artists and designers. Widest such selection available. All illustrations in line. 119pp. 8⅜ x 11¼.
23694-3 Pa. $3.50

WOMEN: A PICTORIAL ARCHIVE FROM NINETEENTH-CENTURY SOURCES, edited by Jim Harter. 391 copyright-free wood engravings for artists and designers selected from rare periodicals. Most extensive such collection available. All illustrations in line. 128pp. 9 x 12.
23703-6 Pa. $4.50

ARABIC ART IN COLOR, Prisse d'Avennes. From the greatest ornamentalists of all time—50 plates in color, rarely seen outside the Near East, rich in suggestion and stimulus. Includes 4 plates on covers. 46pp. 9⅜ x 12¼. 23658-7 Pa. $6.00

AUTHENTIC ALGERIAN CARPET DESIGNS AND MOTIFS, edited by June Beveridge. Algerian carpets are world famous. Dozens of geometrical motifs are charted on grids, color-coded, for weavers, needleworkers, craftsmen, designers. 53 illustrations plus 4 in color. 48pp. 8¼ x 11. (Available in U.S. only) 23650-1 Pa. $1.75

DICTIONARY OF AMERICAN PORTRAITS, edited by Hayward and Blanche Cirker. 4000 important Americans, earliest times to 1905, mostly in clear line. Politicians, writers, soldiers, scientists, inventors, industrialists, Indians, Blacks, women, outlaws, etc. Identificatory information. 756pp. 9¼ x 12¾. 21823-6 Clothbd. $40.00

HOW THE OTHER HALF LIVES, Jacob A. Riis. Journalistic record of filth, degradation, upward drive in New York immigrant slums, shops, around 1900. New edition includes 100 original Riis photos, monuments of early photography. 233pp. 10 x 7⅞. 22012-5 Pa. $7.00

NEW YORK IN THE THIRTIES, Berenice Abbott. Noted photographer's fascinating study of city shows new buildings that have become famous and old sights that have disappeared forever. Insightful commentary. 97 photographs. 97pp. 11⅜ x 10. 22967-X Pa. $5.00

MEN AT WORK, Lewis W. Hine. Famous photographic studies of construction workers, railroad men, factory workers and coal miners. New supplement of 18 photos on Empire State building construction. New introduction by Jonathan L. Doherty. Total of 69 photos. 63pp. 8 x 10¾.
23475-4 Pa. $3.00

THE DEPRESSION YEARS AS PHOTOGRAPHED BY ARTHUR ROTHSTEIN, Arthur Rothstein. First collection devoted entirely to the work of outstanding 1930s photographer: famous dust storm photo, ragged children, unemployed, etc. 120 photographs. Captions. 119pp. 9¼ x 10¾.
23590-4 Pa. $5.00

CAMERA WORK: A PICTORIAL GUIDE, Alfred Stieglitz. All 559 illustrations and plates from the most important periodical in the history of art photography, *Camera Work* (1903-17). Presented four to a page, reduced in size but still clear, in strict chronological order, with complete captions. Three indexes. Glossary. Bibliography. 176pp. 8⅜ x 11¼.
23591-2 Pa. $6.95

ALVIN LANGDON COBURN, PHOTOGRAPHER, Alvin L. Coburn. Revealing autobiography by one of greatest photographers of 20th century gives insider's version of Photo-Secession, plus comments on his own work. 77 photographs by Coburn. Edited by Helmut and Alison Gernsheim. 160pp. 8⅛ x 11. 23685-4 Pa. $6.00

NEW YORK IN THE FORTIES, Andreas Feininger. 162 brilliant photographs by the well-known photographer, formerly with *Life* magazine, show commuters, shoppers, Times Square at night, Harlem nightclub, Lower East Side, etc. Introduction and full captions by John von Hartz. 181pp. 9¼ x 10¾. 23585-8 Pa. $6.00

GREAT NEWS PHOTOS AND THE STORIES BEHIND THEM, John Faber. Dramatic volume of 140 great news photos, 1855 through 1976, and revealing stories behind them, with both historical and technical information. Hindenburg disaster, shooting of Oswald, nomination of Jimmy Carter, etc. 160pp. 8¼ x 11. 23667-6 Pa. $5.00

THE ART OF THE CINEMATOGRAPHER, Leonard Maltin. Survey of American cinematography history and anecdotal interviews with 5 masters—Arthur Miller, Hal Mohr, Hal Rosson, Lucien Ballard, and Conrad Hall. Very large selection of behind-the-scenes production photos. 105 photographs. Filmographies. Index. Originally *Behind the Camera.* 144pp. 8¼ x 11. 23686-2 Pa. $5.00

DESIGNS FOR THE THREE-CORNERED HAT (LE TRICORNE), Pablo Picasso. 32 fabulously rare drawings—including 31 color illustrations of costumes and accessories—for 1919 production of famous ballet. Edited by Parmenia Migel, who has written new introduction. 48pp. 9⅜ x 12¼. (Available in U.S. only) 23709-5 Pa. $5.00

NOTES OF A FILM DIRECTOR, Sergei Eisenstein. Greatest Russian filmmaker explains montage, making of *Alexander Nevsky,* aesthetics; comments on self, associates, great rivals (Chaplin), similar material. 78 illustrations. 240pp. 5⅜ x 8½. 22392-2 Pa. $4.50

PRINCIPLES OF ORCHESTRATION, Nikolay Rimsky-Korsakov. Great classical orchestrator provides fundamentals of tonal resonance, progression of parts, voice and orchestra, tutti effects, much else in major document. 330pp. of musical excerpts. 489pp. 6½ x 9¼. 21266-1 Pa. **$7.50**

TRISTAN UND ISOLDE, Richard Wagner. Full orchestral score with complete instrumentation. Do not confuse with piano reduction. Commentary by Felix Mottl, great Wagnerian conductor and scholar. Study score. 655pp. 8⅛ x 11. 22915-7 Pa. $13.95

REQUIEM IN FULL SCORE, Giuseppe Verdi. Immensely popular with choral groups and music lovers. Republication of edition published by C. F. Peters, Leipzig, n. d. German frontmaker in English translation. Glossary. Text in Latin. Study score. 204pp. 9⅜ x 12¼.
23682-X Pa. $6.00

COMPLETE CHAMBER MUSIC FOR STRINGS, Felix Mendelssohn. All of Mendelssohn's chamber music: Octet, 2 Quintets, 6 Quartets, and Four Pieces for String Quartet. (Nothing with piano is included). Complete works edition (1874-7). Study score. 283 pp. 9⅜ x 12¼.
23679-X Pa. **$7.50**

POPULAR SONGS OF NINETEENTH-CENTURY AMERICA, edited by Richard Jackson. 64 most important songs: "Old Oaken Bucket," "Arkansas Traveler," "Yellow Rose of Texas," etc. Authentic original sheet music, full introduction and commentaries. 290pp. 9 x 12. 23270-0 Pa. **$7.95**

COLLECTED PIANO WORKS, Scott Joplin. Edited by Vera Brodsky Lawrence. Practically all of Joplin's piano works—rags, two-steps, marches, waltzes, etc., 51 works in all. Extensive introduction by Rudi Blesh. Total of 345pp. 9 x 12. 23106-2 Pa. $14.95

BASIC PRINCIPLES OF CLASSICAL BALLET, Agrippina Vaganova. Great Russian theoretician, teacher explains methods for teaching classical ballet; incorporates best from French, Italian, Russian schools. 118 illustrations. 175pp. 5⅜ x 8½. 22036-2 Pa. $2.50

CHINESE CHARACTERS, L. Wieger. Rich analysis of 2300 characters according to traditional systems into primitives. Historical-semantic analysis to phonetics (Classical Mandarin) and radicals. 820pp. 6⅛ x 9¼.
21321-8 Pa. $10.00

EGYPTIAN LANGUAGE: EASY LESSONS IN EGYPTIAN HIEROGLYPHICS, E. A. Wallis Budge. Foremost Egyptologist offers Egyptian grammar, explanation of hieroglyphics, many reading texts, dictionary of symbols. 246pp. 5 x 7½. (Available in U.S. only)
21394-3 Clothbd. $7.50

AN ETYMOLOGICAL DICTIONARY OF MODERN ENGLISH, Ernest Weekley. Richest, fullest work, by foremost British lexicographer. Detailed word histories. Inexhaustible. Do not confuse this with *Concise Etymological Dictionary,* which is abridged. Total of 856pp. 6½ x 9¼.
21873-2, 21874-0 Pa., Two-vol. set $12.00

UNCLE SILAS, J. Sheridan LeFanu. Victorian Gothic mystery novel, considered by many best of period, even better than Collins or Dickens. Wonderful psychological terror. Introduction by Frederick Shroyer. 436pp. 5⅜ x 8½. 21715-9 Pa. $6.00

JURGEN, James Branch Cabell. The great erotic fantasy of the 1920's that delighted thousands, shocked thousands more. Full final text, Lane edition with 13 plates by Frank Pape. 346pp. 5⅜ x 8½. 23507-6 Pa. $4.50

THE CLAVERINGS, Anthony Trollope. Major novel, chronicling aspects of British Victorian society, personalities. Reprint of Cornhill serialization, 16 plates by M. Edwards; first reprint of full text. Introduction by Norman Donaldson. 412pp. 5⅜ x 8½. 23464-9 Pa. $5.00

KEPT IN THE DARK, Anthony Trollope. Unusual short novel about Victorian morality and abnormal psychology by the great English author. Probably the first American publication. Frontispiece by Sir John Millais. 92pp. 6½ x 9¼. 23609-9 Pa. $2.50

RALPH THE HEIR, Anthony Trollope. Forgotten tale of illegitimacy, inheritance. Master novel of Trollope's later years. Victorian country estates, clubs, Parliament, fox hunting, world of fully realized characters. Reprint of 1871 edition. 12 illustrations by F. A. Faser. 434pp. of text. 5⅜ x 8½. 23642-0 Pa. $5.00

YEKL and THE IMPORTED BRIDEGROOM AND OTHER STORIES OF THE NEW YORK GHETTO, Abraham Cahan. Film *Hester Street* based on *Yekl* (1896). Novel, other stories among first about Jewish immigrants of N.Y.'s East Side. Highly praised by W. D. Howells—Cahan "a new star of realism." New introduction by Bernard G. Richards. 240pp. 5⅜ x 8½. 22427-9 Pa. $3.50

THE HIGH PLACE, James Branch Cabell. Great fantasy writer's enchanting comedy of disenchantment set in 18th-century France. Considered by some critics to be even better than his famous *Jurgen.* 10 illustrations and numerous vignettes by noted fantasy artist Frank C. Pape. 320pp. 5⅜ x 8½. 23670-6 Pa. $4.00

ALICE'S ADVENTURES UNDER GROUND, Lewis Carroll. Facsimile of ms. Carroll gave Alice Liddell in 1864. Different in many ways from final Alice. Handlettered, illustrated by Carroll. Introduction by Martin Gardner. 128pp. 5⅜ x 8½. 21482-6 Pa. $2.00

FAVORITE ANDREW LANG FAIRY TALE BOOKS IN MANY COLORS, Andrew Lang. The four Lang favorites in a boxed set—the complete *Red, Green, Yellow* and *Blue* Fairy Books. 164 stories; 439 illustrations by Lancelot Speed, Henry Ford and G. P. Jacomb Hood. Total of about 1500pp. 5⅜ x 8½. 23407-X Boxed set, Pa. $14.95

HOUSEHOLD STORIES BY THE BROTHERS GRIMM. All the great Grimm stories: "Rumpelstiltskin," "Snow White," "Hansel and Gretel," etc., with 114 illustrations by Walter Crane. 269pp. 5⅜ x 8½.
21080-4 Pa. $3.50

SLEEPING BEAUTY, illustrated by Arthur Rackham. Perhaps the fullest, most delightful version ever, told by C. S. Evans. Rackham's best work. 49 illustrations. 110pp. 7⅞ x 10¾. 22756-1 Pa. $2.50

AMERICAN FAIRY TALES, L. Frank Baum. Young cowboy lassoes Father Time; dummy in Mr. Floman's department store window comes to life; and 10 other fairy tales. 41 illustrations by N. P. Hall, Harry Kennedy, Ike Morgan, and Ralph Gardner. 209pp. 5⅜ x 8½. 23643-9 Pa. $3.00

THE WONDERFUL WIZARD OF OZ, L. Frank Baum. Facsimile in full color of America's finest children's classic. Introduction by Martin Gardner. 143 illustrations by W. W. Denslow. 267pp. 5⅜ x 8½.
20691-2 Pa. $3.50

THE TALE OF PETER RABBIT, Beatrix Potter. The inimitable Peter's terrifying adventure in Mr. McGregor's garden, with all 27 wonderful, full-color Potter illustrations. 55pp. 4¼ x 5½. (Available in U.S. only)
22827-4 Pa. $1.25

THE STORY OF KING ARTHUR AND HIS KNIGHTS, Howard Pyle. Finest children's version of life of King Arthur. 48 illustrations by Pyle. 131pp. 6⅛ x 9¼. 21445-1 Pa. $4.95

CARUSO'S CARICATURES, Enrico Caruso. Great tenor's remarkable caricatures of self, fellow musicians, composers, others. Toscanini, Puccini, Farrar, etc. Impish, cutting, insightful. 473 illustrations. Preface by M. Sisca. 217pp. 8⅜ x 11¼. 23528-9 Pa. $6.95

PERSONAL NARRATIVE OF A PILGRIMAGE TO ALMADINAH AND MECCAH, Richard Burton. Great travel classic by remarkably colorful personality. Burton, disguised as a Moroccan, visited sacred shrines of Islam, narrowly escaping death. Wonderful observations of Islamic life, customs, personalities. 47 illustrations. Total of 959pp. 5⅜ x 8½.
21217-3, 21218-1 Pa., Two-vol. set $12.00

INCIDENTS OF TRAVEL IN YUCATAN, John L. Stephens. Classic (1843) exploration of jungles of Yucatan, looking for evidences of Maya civilization. Travel adventures, Mexican and Indian culture, etc. Total of 669pp. 5⅜ x 8½. 20926-1, 20927-X Pa., Two-vol. set $7.90

AMERICAN LITERARY AUTOGRAPHS FROM WASHINGTON IRVING TO HENRY JAMES, Herbert Cahoon, et al. Letters, poems, manuscripts of Hawthorne, Thoreau, Twain, Alcott, Whitman, 67 other prominent American authors. Reproductions, full transcripts and commentary. Plus checklist of all American Literary Autographs in The Pierpont Morgan Library. Printed on exceptionally high-quality paper. 136 illustrations. 212pp. 9⅛ x 12¼. 23548-3 Pa. $12.50

AN AUTOBIOGRAPHY, Margaret Sanger. Exciting personal account of hard-fought battle for woman's right to birth control, against prejudice, church, law. Foremost feminist document. 504pp. 5⅜ x 8½.
20470-7 Pa. $5.50

MY BONDAGE AND MY FREEDOM, Frederick Douglass. Born as a slave, Douglass became outspoken force in antislavery movement. The best of Douglass's autobiographies. Graphic description of slave life. Introduction by P. Foner. 464pp. 5⅜ x 8½. 22457-0 Pa. $5.50

LIVING MY LIFE, Emma Goldman. Candid, no holds barred account by foremost American anarchist: her own life, anarchist movement, famous contemporaries, ideas and their impact. Struggles and confrontations in America, plus deportation to U.S.S.R. Shocking inside account of persecution of anarchists under Lenin. 13 plates. Total of 944pp. 5⅜ x 8½.
22543-7, 22544-5 Pa., Two-vol. set $12.00

LETTERS AND NOTES ON THE MANNERS, CUSTOMS AND CONDITIONS OF THE NORTH AMERICAN INDIANS, George Catlin. Classic account of life among Plains Indians: ceremonies, hunt, warfare, etc. Dover edition reproduces for first time all original paintings. 312 plates. 572pp. of text. 6⅛ x 9¼. 22118-0, 22119-9 Pa.. Two-vol. set $12.00

THE MAYA AND THEIR NEIGHBORS, edited by Clarence L. Hay, others. Synoptic view of Maya civilization in broadest sense, together with Northern, Southern neighbors. Integrates much background, valuable detail not elsewhere. Prepared by greatest scholars: Kroeber, Morley, Thompson, Spinden, Vaillant, many others. Sometimes called Tozzer Memorial Volume. 60 illustrations, linguistic map. 634pp. 5⅜ x 8½.
23510-6 Pa. $7.50

HANDBOOK OF THE INDIANS OF CALIFORNIA, A. L. Kroeber. Foremost American anthropologist offers complete ethnographic study of each group. Monumental classic. 459 illustrations, maps. 995pp. 5⅜ x 8½.
23368-5 Pa. $13.00

SHAKTI AND SHAKTA, Arthur Avalon. First book to give clear, cohesive analysis of Shakta doctrine, Shakta ritual and Kundalini Shakti (yoga). Important work by one of world's foremost students of Shaktic and Tantric thought. 732pp. 5⅜ x 8½. (Available in U.S. only)
23645-5 Pa. $7.95

AN INTRODUCTION TO THE STUDY OF THE MAYA HIEROGLYPHS, Syvanus Griswold Morley. Classic study by one of the truly great figures in hieroglyph research. Still the best introduction for the student for reading Maya hieroglyphs. New introduction by J. Eric S. Thompson. 117 illustrations. 284pp. 5⅜ x 8½. 23108-9 Pa. $4.00

A STUDY OF MAYA ART, Herbert J. Spinden. Landmark classic interprets Maya symbolism, estimates styles, covers ceramics, architecture, murals, stone carvings as artforms. Still a basic book in area. New introduction by J. Eric Thompson. Over 750 illustrations. 341pp. 8⅜ x 11¼.
21235-1 Pa. $6.95